Lecture Notes in Mathematics

Edited by A. Dold and B. Eckmann

630

Numerical Analysis

Proceedings of the Biennial Conference
Held at Dundee, June 28 – July 1, 1977

Edited by G. A. Watson

Springer-Verlag
Berlin Heidelberg New York 1978

Editor
G. A. Watson
University of Dundee
Department of Mathematics
Dundee, DD1 4HN/Scotland

AMS Subject Classifications (1970): 65-02, 65D10, 65F05, 65F20, 65K05, 65L05, 65L10, 65M99, 65N30, 65R05

ISBN 3-540-08538-6 Springer-Verlag Berlin Heidelberg New York
ISBN 0-387-08538-6 Springer-Verlag New York Heidelberg Berlin

Printed in Germany

Printing and binding: Beltz Offsetdruck, Hemsbach/Bergstr.
2141/3140-543210

Preface

For the 4 days June 28 - July 1, 1977, over 220 people attended the 7th Dundee Biennial Conference on Numerical Analysis at the University of Dundee, Scotland. The technical program consisted of 16 invited papers, and 63 short submitted papers, the contributed talks being given in 3 parallel sessions. This volume contains, in complete form, the papers given by the invited speakers, and a list of all other papers presented.

I would like to take this opportunity of thanking the speakers, including the after dinner speaker at the conference dinner, Professor D S Jones, all chairmen and participants for their contributions. I would also like to thank the many people in the Mathematics Department of this University who assisted in various ways with the preparation for, and running of, this conference. In particular, the considerable task of typing the various documents associated with the conference, and some of the typing in this volume has been done by Miss R Dudgeon; this work is gratefully acknowledged.

G A Watson

Dundee, September 1977.

CONTENTS

INVITED SPEAKERS

C T H Baker	Department of Mathematics, University of Manchester, Oxford Road, Manchester M13 9PL, England.
I Barrodale	Department of Mathematics, University of Victoria, P.O. Box 1700, Victoria, B.C., Canada.
J S R Chisholm	Mathematical Institute, The University, Canterbury, Kent CT2 7NF, England.
L Collatz	Institut für Angewandte Mathematik, Universitat Hamburg, 2 Hamburg 13, Bundesstr 55, W Germany.
M G Cox	Division of Numerical Analysis and Computing, National Physical Laboratory, Teddington, Middlesex TW11 OLW, England.
J Douglas, Jnr	Department of Mathematics, The University of Chicago, 5734 University Avenue, Chicago, Illinois 60637, USA.
J A George	Department of Computer Science, University of Waterloo, Ontario, Canada.
G H Golub	Computer Science Department, Stanford University, Stanford, California 94305, USA.
D S Jones	Department of Mathematics, University of Dundee, Dundee DD1 4HN, Scotland.
A R Mitchell	Department of Mathematics, University of Dundee, Dundee DD1 4HN, Scotland.
J J Moré	Applied Mathematics Division, Argonne National Laboratory, 9700 South Cass Avenue, Argonne, Illinois 60439, USA.
M R Osborne	Computer Centre, Australian National University, Box 4 P.O., Canberra, A.C.T. 2600, Australia.
V Pereyra	Applied Mathematics 101-50, California Institute of Technology, Pasadena, California 91125, USA.
M J D Powell	Department of Applied Mathematics and Theoretical Physics, University of Cambridge, Silver Street, Cambridge CB3 9EW, England.
R W H Sargent	Department of Chemical Engineering and Chemical Technology, Imperial College, London SW7, England.
H J Stetter	Institut für Numerische Mathematik, Technische Hochschule Wien, A-1040 Wien, Gusshausstr, 27-29 Austria.
E L Wachspress	General Electric Company, P.O. Box 1072, Schenectady, New York 12301, USA.

SUBMITTED PAPERS

Z Aktas: Computer Science Dept, Middle East Technical University, Turkey.
An accuracy improvement for the method of lines.

P Alfeld: Mathematics Dept, University of Dundee, Scotland.
CDS - A new technique for certain stiff systems of ordinary differential equations.

K Balla: Computer and Automation Institute, Hungarian Academy of Science.
On error estimates of the substitution of the boundedness condition on solutions of systems of linear ordinary differential equations with regular singularity.

K E Barrett: Mathematics Dept, Lanchester Polytechnic, England.
The finite integral method for partial differential equations.

D G Bettis: Institute for Mathematics, Technical University of Munich, Germany.
An efficient embedded Runge-Kutta method.

Jean Beuneu: University of Lille I, France.
The rebalancing method for solving linear systems and eigenproblems.

Ake Björck: Mathematics Dept, Linköping University, Sweden.
Iterative solution of under- and overdetermined linear systems.

Klaus W A Böhmer: Mathematics Institute, University of Karlsruhe, Germany.
Defect corrections via neighbouring problems.

Claude Brezinski: University of Lille I, France.
Rational approximants to power series.

Hermann Brunner: Mathematics Department, Dalhousie University, Canada.
Volterra integral equations and their discretizations.

J P Coleman: Mathematics Dept, University of Durham, England.
Evaluation of the Bessel Functions J_0 and J_1 of complex argument.

I D Coope: Mathematics Dept, University of Dundee, Scotland.
Global convergence results for augmented Lagrangian methods.

C J Cooper: School of Math. and Physical Science, University of Sussex, England.
The order of convergence of linear methods for ordinary differential equations.

L J Cromme: Mathematics Dept, University of Bonn, Germany.
Numerical methods for nonlinear maximum norm approximations.

L M Delves: Dept of Comp and Statistical Science, University of Liverpool, England.
A global element method for the solution of elliptic partial differential equations.

P M Dew: Centre for Computer Studies, University of Leeds, England.
Numerical solution of quasi-linear heat problems with error estimates.

I S Duff: A.E.R.E. Harwell, England.
MA28 - a set of subroutines for solving sparse unsymmetric linear equations.

S Ellacott: Mathematics Dept, Brighton Polytechnic, England.
Practical complex best approximation: The state of the art.

C M Elliott: Computing Laboratory, Oxford University, England.
On the numerical solution of an electrochemical machining problem via a variational inequality formulation.

G Elliott: Mathematics Dept, Portsmouth Polytechnic, England.
The construction of Chebyshev approximations in the complex plane.

R England and J P Hennart: Universidad Nacional de Mexico.
Fractional steps finite element techniques for strongly anisotropic diffusion problems.

R Fletcher: Mathematics Dept, University of Dundee, Scotland.
The reduced Hessian in variable metric methods.

T L Freeman: Mathematics Dept., University of Manchester, England.
A method for computing the zeros of a polynomial with real coefficients.

Nima Geffen and Sara Yaniv: Tel-Aviv University, Israel.
Isoparametric characteristic elements for the Tricomi equation.

B Germain-Bonne, University of Lille I, France.
Shape and variation diminishing properties of spline curves.

Michael Ghil and Remesh Balgovind: Courant Institute of Mathematical Sciences, New York University, USA.
A fast Cauchy-Riemann solver with nonlinear applications.

Ian Gladwell: Mathematics Dept, University of Manchester, England.
The NAG library chapter for the solution of ordinary differential equations.

Moshe Goldberg: Mathematics Dept, University of California, USA.
Dissipative schemes for hyperbolic problems and boundary extrapolation.

R Gorenflo: Mathematics Dept, Freie Universität Berlin, Germany.
Conservative difference schemes for diffusion problems.

Myron S Henry: Mathematics Dept, Montana State University, USA.
Numerical comparisons of algorithms for polynomial and rational multivariate approximations.

J N Holt: Mathematics Dept, University of Queensland, Australia.
Free-knot cubic spline inversion of a Fredholm integral equation.

M K Horn: Institute for Mathematics, Technical University of Munich, Germany.
Developments in high-order Runge-Kutta-Nyström methods.

W D Hoskins, D S Meek, D J Walton: Dept of Computer Science, University of Manitoba, Canada.
An alternative method for the solution of Poisson-type equations on Rectangular Regions in two or three space dimensions.

K Jittorntrum, M R Osborne: Computer Centre, Australian National University.
Trajectory analysis and extrapolation in barrier function methods.

D C Joyce: Mathematics Dept, Massey University, New Zealand.
Extrapolation to the limit - algorithms and applications.

Bo Kagström: Dept of Information Processing and Numerical Analysis, University of Umea, Sweden.
On the numerical computation of matrix functions.

Malcolm S Keech: Mathematics Dept, University of Manchester, England.
Semi-explicit methods in the numerical solution of first kind Volterra integral equations.

R Kress: Universität Göttingen and University of Strathclyde, Scotland.
On improving the rate of convergence of successive approximation for integral equations of potential theory.

D P Laurie: National Research Institute for Mathematical Sciences, South Africa.
Exponentially fitted multipoint methods for two-point boundary value problems.

J D Lawson and J Ll Morris: Computer Science Dept, University of Waterloo, Canada.
The extrapolation of first order methods for parabolic partial differential equations.

A V Levy*and A Montalvo+: *Universidad Nacional Autónoma de México, +Universidad Iberoamericana, México.
The tunneling algorithm for the global minimization of functions.

I M Longman: Dept of Geophysics and Planetary Sciences, Tel-Aviv University, Israel.
A method of Laplace transform inversion by exponential series.

Jens Lorenz: Institute for Numerical Mathematics, University of Munster, Germany.
Stability inequalities for discrete boundary value problems.

J T Marti: Mathematics Dept, Swiss Federal Institute of Technology.
An algorithm for the computation of Fourier coefficients of non-analytic functions using B-splines of arbitrary order.

J C Mason: Mathematics Branch, Royal Military College of Science, Shrivenham, England.
A one-dimensional spline approximation method for the numerical solution of heat conduction problems.

S McKee: University of Oxford, England.
Multistep methods for solving linear Volterra integro-differential equations.

G Moore and A Spence: School of Mathematics, University of Bath, England.
Newton's method near a bifurcation point.

P Moore: Mathematics Dept, University of Aston in Birmingham, England.
Finite element multistep multiderivative schemes for linear parabolic equations.

Gerhard Opfer: Mathematics Dept, University of Hamburg, Germany.
Numerical solution of certain nonstandard approximation problems.

I Riddell: Dept of Computational and Statistical Science, University of Liverpool, England.
On comparing integral equation routines.

A Robinson and A Prothero: Shell Research Limited, Chester, England.
Global error estimates for solutions to stiff systems of ordinary differential equations.

J Barkley Rosser: Mathematics Research Center, University of Wisconsin-Madison, USA.
Harmonic functions on regions with reentrant corners.

A Sayfy: School of Maths. and Physical Sciences, University of Sussex, England.
Additive numerical methods for ordinary differential equations.

J Sinclair: Mathematics Dept, University of Dundee, Scotland.
A variable metric method generating orthogonal directions.

H J J te Riele: Mathematical Centre, Amsterdam, Holland.
Computation of zeros of partial sums of the Riemann ζ-function with real part > 1.

Per Grove Thomsen and Zahari Zlatev: Institute for Numerical Analysis, Technical University of Denmark.
The use of Backward Differentiation methods in the solution of non-stationary heat conduction problems.

Ph L Toint: F.N.D.P. Belguim.
On sparse and symmetric matrix updating subject to a linear equation.

J M Watt: Dept of Computational Science, University of Liverpool, England.
The convergence of deferred and defect corrections.

Richard Weiss: Technische Universität, Wien, Austria.
On the eigenvalue problem for singular systems of ordinary differential equations.

B Werner: Mathematics Dept., University of Hamburg, Germany.
About a connection between complementary and nonconforming finite elements.

Ragnar Winther: Institute of Informatics, University of Oslo, Norway.
A Galerkin method for a parabolic control problem.

G Woodford: Mathematics Dept, University of Dundee, Scotland.
Isoparametric cubic triangles in the finite element method.

K Wright: Computing Laboratory, University of Newcastle upon Tyne, England.
Asymptotic properties of quadrature weights based on zeros of orthogonal polynomials over partial and full ranges.

RUNGE-KUTTA METHODS FOR VOLTERRA INTEGRAL EQUATIONS OF THE SECOND KIND

Christopher T.H. Baker

Abstract. *Some Runge-Kutta methods for the numerical solution of Volterra integral equations of the second kind have been considered previously, and these methods can be generalized in a natural way. By considering a class of variable-step quadrature methods, the Runge-Kutta methods appear as extensions of the step-by-step quadrature methods, and theoretical insight is readily obtained. Such insight may provide confidence limits when constructing practical algorithms.*

1. Introduction

There is a growing literature (see the references in Baker [1], for example) devoted to the numerical treatment of integral equations of the form

$$f(x) - \int_0^x F(x, y; f(y))dy = g(x) \qquad (X \geq x \geq 0). \tag{1.1}$$

In eqn (1.1), we assume continuity of $g(x)$ for $x \in [0, X]$ and of $F(x, y; v)$ for $0 \leq y \leq x \leq X$, $|v| < \infty$, where X is some fixed value; we further assume a Lipschitz condition $|F(x, y; v_1) - F(x, y; v_2)| \leq L|v_1 - v_2|$ uniformly for $0 \leq y \leq x \leq X$, $|v_1|, |v_2| < \infty$. Then eqn (1.1) has a unique solution $f(x) \in C[0, X]$; the equation itself is known as a Volterra equation of the second kind.

A standard *quadrature method* for the approximate solution of eqn (1.1) involves a choice of stepsize h and a sequence of quadrature rules

$$\int_0^{ih} \phi(y)dy \simeq h \sum_{k=0}^{i} w_{i,k}\phi(kh) \qquad (i = 1, 2, \ldots, N; \ Nh=X) \tag{1.2}$$

associated with a tableau of weights

$$\begin{array}{lllll} w_{0,0} & & & & \\ w_{1,0} & w_{1,1} & & & \\ w_{2,0} & w_{2,1} & w_{2,2} & & \\ w_{3,0} & w_{3,1} & w_{3,2} & w_{3,3} & \\ \cdot & \cdot & \cdot & \cdot & \cdot \\ \cdot & \cdot & \cdot & \cdot & \cdot\ \cdot \\ \cdot & \cdot & \cdot & \cdot & \cdot\ \cdot\ \cdot\ , \end{array} \tag{1.3}$$

with $w_{0,0} = 0$. When used to discretize eqn (1.1), the rules (1.2) yield the system of equations

$$\tilde{f}(ih) - h \sum_{k=0}^{i} w_{i,k} F(ih, kh; \tilde{f}(kh)) = g(ih) \quad (i = 1,2,\ldots,N) \tag{1.4}$$

which with the equation $\tilde{f}(0) = f(0) = g(0)$ determine values $\tilde{f}(ih)$ approximating $f(ih)$ for $i = 0, 1, \ldots, N$. With certain choices of the rules (1.2), equations (1.4) reduce, in the case where $F(x, y; v) = \Phi(y, v)$ and $g(x)$ is constant, to linear multistep or cyclic linear multistep methods for the initial value problem

$$f'(x) = \Phi(x, f(x)), \tag{1.5}$$

with $f(0) = g(0)$. Thus, Gregory rules of fixed order in (1.2) yield Adams methods, given suitable starting values.

A convergence theory for (1.4) is given by Linz [14] and an asymptotic stability analysis (paralleling the work of Henrici [10] for eqn (1.5)) is given [11] by Kobayasi. The work of the latter author and the work of Linz on stability theory has been illuminated by Noble [16], who emphasises the importance§ of the repetition factor of the weights $\{w_{\mu,\nu}\}$ defined as follows:

Definition 1.1 A tableau of weights $\{w_{\mu,\nu}\}$ of the form (1.3) has (row-) repetition factor m if m is the smallest integer such that $w_{m+\mu,\nu} = w_{\mu,\nu}$ for $q_1 \leqslant \nu \leqslant \mu - q_2$ where q_1, q_2 are independent of μ.

2. Runge-Kutta methods

Pouzet [17] and Beltyukov [5] have both considered extensions of Runge-Kutta methods for (1.5) to the treatment of eqn (1.1), whilst Weiss [18] defines methods which he calls "implicit Runge-Kutta methods". Baker [1] gives an introduction to the work of these authors, and the work of Garey [8] is also relevant.

We shall here restrict attention to simple extensions of methods of the type considered by Pouzet for a canonical form of eqn (1.1), namely $f(x) = \int_0^x G(x, y; f(y))dy$. Pouzet employs the parameters $\{A_{r,t}, \theta_t\}$ of a p-stage explicit Runge-Kutta formula for (1.5) [Lambert, 11].

The parameters of a class of Runge-Kutta formulae can be displayed conveniently in tabular form as:

$$\begin{array}{c|cccccc}
\theta_0 & A_{0,0} & \cdot & \cdot & A_{0,p-1} & A_{0,p} \\
\theta_1 & A_{1,0} & A_{1,1} & \cdot & A_{1,p-1} & A_{1,p} \\
\theta_2 & A_{2,0} & A_{2,1} & A_{2,2} & A_{2,p-1} & A_{2,p} \\
\cdot & \cdot & \cdot & \cdot & & \\
\cdot & \cdot & \cdot & \cdot & & \\
\theta_{p-1} & A_{p-1,0} & A_{p-1,1} & \cdots & A_{p-1,p-1} & A_{p-1,p} \\
\hline
\theta_p = 1 & A_{p,0} & A_{p,1} & \cdots & A_{p,p-1} & A_{p,p}
\end{array} \quad . \tag{2.1}$$

§ Linz proposes the hypothesis that if the tableau of weights associated with (1.2) has a repetition factor greater than unity then the scheme (1.4) has undesirable stability properties. The results of Baker and Keech [2, 4] give additional credence to this hypothesis. We return to this topic later.

Butcher [6], and Lapidus and Seinfeld [13], use a similar notation but omit the last column. The Runge-Kutta formula associated with (2.1) is *explicit* if $A_{r,t} = 0$ for $t \geq r$ and *semi-explicit* if $A_{r,t} = 0$ for $t > r$, $r = 0, 1, \ldots, p$; otherwise it is (formally) *implicit*.†

For some gain in simplicity, we shall assume (except, as in §5, otherwise stated) that *the Runge-Kutta formula is explicit or semi-explicit*, the former term being subsumed in the latter, and (see [11]) that

$$\sum_{t=0}^{r} A_{r,t} = \theta_r \qquad (r = 0, 1, \ldots, p). \tag{2.2}$$

Provided that $\theta_p = 1$ we shall regard a semi-explicit formula as '*feasible*', and we shall call a feasible formula '*acceptable*' if $\theta_r \in [0, 1]$ for $r = 0,1,\ldots, p$ and '*convenient*' if $0 \leq \theta_0 \leq \theta_1 \leq \ldots \leq \theta_p = 1$. (Pouzet [17] discusses only the case where the formula is explicit and convenient.) We will also refer to a semi-explicit formula as an '*equally-spacing Runge-Kutta formula*' in the case $\theta_r = r/p$, $r = 0,1, \ldots, p$.

The tableau (2.1) is associated with a set of quadrature formulae

$$\int_0^{\theta_r h} \phi(y)dy \simeq h \sum_{t=0}^{r} A_{r,t}\ \phi(\theta_t h) \qquad (r = 0, 1, \ldots, p) \tag{2.3}$$

each of which is exact if $\phi(x)$ is constant for $x \in [0, \theta_r h]$, by virtue of (2.2). In view of (2.3) we are tempted to modify the equations (1.4) and seek approximations $\tilde{f}_j$ to $f(ih + \theta_r h)$ where, denoting the integer part of z by $[z]$,

$$r \equiv (j-1) \bmod (p+1), \ i = [(j-1)/(p+1)], \quad j = 1, 2, 3, \ldots \tag{2.4}$$

$(i = 0, 1, \ldots, N-1;\ r = 0, 1, \ldots, p)$. We employ the notation $\tilde{f}_j$ rather than $\tilde{f}(ih + \theta_r h)$ to avoid confusion if $\theta_r = \theta_\rho$ for $r \neq \rho$. (Similarly, $\tilde{f}_1$ denotes an approximation to $f(\theta_0 h)$ even if $\theta_0 = 0$.) We will write $y_j = ih + \theta_r h$ where i, j, r are related by (2.4). We set $x = y_j$ in eqn (1.1) and discretize with the quadrature rules

$$\int_0^{y_j} \phi(y)dy \simeq h \sum_{k=0}^{j} \Omega_{j,k}\ \phi(y_k) \qquad (j = 1, 2, 3, \ldots) \tag{2.5}$$

to obtain

$$\tilde{f}_j - h \sum_{k=0}^{j} \Omega_{j,k}\ F(y_j, y_k, \tilde{f}_k) = g(y_j) \qquad (j = 1, 2, 3, \ldots) \tag{2.6}$$

---

† It is conventional to omit the supradiagonal zero elements of a semi-explicit formula (and to take $A_{p,p} = 0$ in such a formula). The last column of (2.1) will be taken to consist of zeros if it is omitted (as is usual) from the tableau. We require the last column to include what are effectively block-by-block methods in §5, but the traditional shape of the array can be restored, formally, by including a value $\theta_{p+1} = 1$, repeating the final row of (2.1).

along with the prescribed value $\tilde{f}_0 = f(0) = g(0)$. Clearly, the above equations generalize those of (1.4).

Determining the weights $\Omega_{j,k}$ from the Runge-Kutta parameters $A_{r,t}$ of (2.3), possibly in conjunction with a set of weights $\{w_{\mu,\nu}\}$ of (1.2), yields various Runge-Kutta methods. Thus we may set $\Omega_{j,0} = 0$ and otherwise, using a semi-explicit Runge-Kutta formula,

$$\begin{aligned} \Omega_{j,k} &= A_{p,s} \quad \text{for} \quad k \leqslant i(p+1) \\ \Omega_{j,k} &= A_{r,s} \quad \text{for} \quad k > i(p+1) \end{aligned} \tag{2.7}$$

where

$$s \equiv (k-1) \bmod (p+1) \tag{2.8}$$

and where r, i are defined by (2.4). Alternatively we may use (2.7) for small j (say $j \leqslant j_0$ for some j_0) whilst for $j > j_0$ we set, using the weights of (1.2), $\Omega_{j,0} = w_{i,0}$ and for $k \neq 0$

$$\begin{aligned} \Omega_{j,k} &= 0 \quad \text{if} \quad k \leqslant i(p+1) \quad \text{and} \quad s \neq p \\ \Omega_{j,k} &= w_{i,m} \quad \text{if} \quad k \leqslant i(p+1), \quad s = p; \quad m = k/(p+1), \\ \Omega_{j,k} &= A_{r,s} \quad \text{for} \quad k > i(p+1). \end{aligned} \tag{2.9}$$

(Values of s, r and i are given by (2.4), (2.8), as above. Other choices of weights $\Omega_{j,k}$ are of course possible.) We shall refer to the scheme defined by (2.6) with the choice (2.7) as an '*extended Runge-Kutta method*' whilst the choice (2.9) supplies an example of a '*mixed Runge-Kutta method*'. If we replace $F(x, y; v)$ by $G(x, y; v)$ and set $g(x)$ to zero, the extended and mixed Runge-Kutta methods are amongst those described by Pouzet [17] for the canonical form of (1.1), in terms of 'convenient' Runge-Kutta formulae. When the Runge-Kutta formula is acceptable (but not convenient) solving (2.6) requires the evaluation of $F(x, y; v)$ using a value of y which is greater than the corresponding value of x, where $0 \leqslant x \leqslant X$. If the formula is only feasible, then values with $x > X$ and/or $x < 0$ may also be required.

Example 2.1 (a) Consider the third-order Runge-Kutta explicit formula associated with the tableau indicated:

0	0			
$\frac{1}{3}$	$\frac{1}{3}$	0		
$\frac{2}{3}$	0	$\frac{2}{3}$	0	
1	$\frac{1}{4}$	0	$\frac{3}{4}$	0

The corresponding formulae (2.5) are (i) a rectangle (Euler) rule, (ii) the mid-point rule, (iii) a Radau rule. The formula is 'convenient'. We set $\eta = \frac{1}{3}h$ and the extended Runge-Kutta method is then equivalent to solving the equations indicated:

$\tilde{f}(0) = g(0)$, $\tilde{f}(\eta) = g(\eta) + \eta\, F(\eta, 0; \tilde{f}(0))$,
$\tilde{f}(2\eta) = g(2\eta) + 2\eta\, F(2\eta, \eta; \tilde{f}(\eta))$, $\tilde{f}(3\eta) = g(3\eta) + \frac{1}{4}\eta\{9F(3\eta, 2\eta; \tilde{f}(2\eta)) + 3\, F(3\eta, 0; \tilde{f}(0))\}$,
$\tilde{f}(4\eta) = g(4\eta) + \frac{1}{4}\eta\{9F(4\eta, 2\eta; \tilde{f}(2\eta)) + 3F(4\eta, 0; \tilde{f}(0))\} + \eta\, F(4\eta, 3\eta; \tilde{f}(3\eta))$,
etc. and (like all such schemes based on 'equally-spacing' Runge-Kutta formulae) the method reduces to an equal-step quadrature method with, here, h replaced by η.

(b) The Runge-Kutta method associated with the tableau:

$$\begin{array}{c|ccc} \theta_0 & \theta_0 & & \\ \theta_1 & -\sqrt{3}/3 & \theta_0 & \\ \hline \theta_2 = 1 & \frac{1}{2} & \frac{1}{2} & 0 \end{array},$$

where $\theta_0 = (3 + \sqrt{3})/6$, $\theta_1 = (3 - \sqrt{3}/6$, is semi-explicit and 'acceptable' (but not 'convenient'). The first few equations of the extended Runge-Kutta method are of the form:

$(\tilde{f}(0) = g(0)\ )$,
$\tilde{f}(\theta_0 h) = g(\theta_0 h) + \theta_0 h\, F(\theta_0 h, \theta_0 h; \tilde{f}(\theta_0 h))$,
$\tilde{f}(\theta_1 h) = g(\theta_1 h) + \theta_0 h\, F(\theta_1 h, \theta_1 h; \tilde{f}(\theta_1 h))$
$\qquad - (\sqrt{3}/3)h\, F(\theta_1 h, \theta_0 h; \tilde{f}(\theta_0 h))$,
$\tilde{f}(h) = g(h) + \frac{1}{2}h\{F(h, \theta_0 h; \tilde{f}(\theta_0 h)) + F(h, \theta_1 h; \tilde{f}(\theta_1 h))\}$
$\tilde{f}(h + \theta_0 h) = g(h + \theta_0 h) + \frac{1}{2}h\{F(h + \theta_0 h, \theta_0 h; \tilde{f}(\theta_0 h)) + F(h + \theta_0 h, \theta_1 h; \tilde{f}(\theta_1 h)) +$
$\qquad + \theta_0 h\, F(h + \theta_0 h, h + \theta_0 h; \tilde{f}(h + \theta_0 h))$,
etc., and it will be noted that $\theta_0 > \theta_1$.

3. Convergence

Under mild conditions, the approximate values computed using an extended or mixed Runge-Kutta method converge to the true values as the stepsize h tends to zero. Unfortunately, the conditions stated in section 1 for the solvability of eqn (1.1) are not sufficient unless the Runge-Kutta formula is 'convenient'. We shall assume the following condition, which is adequate if we consider 'acceptable' formulae - a refinement is possible for the study of the more general 'feasible' formulae. We shall restrict attention to acceptable formulae.

Condition 3.1 We assume that there exists $\delta > 0$ such that $F(x, y; v)$ is continuous for $0 \leq y \leq x+\delta$, $0 \leq x \leq X$ and $|v| < \infty$, whilst there exists L independent of x, y (with $0 \leq y \leq x+\delta$ and $0 \leq x \leq X$) such that

$$|F(x, y; v_1) - F(x, y; v_2)| \leq L|v_1 - v_2|; \tag{3.1}$$

further, that $g(x) \in C[0, X]$.

Lemma 3.1 Let $\delta > 0$ be given by Condition 3.1. Then $\{\tilde{f}_j | i \leq N-1\}$ is defined by eqn (2.6) for an extended or mixed Runge-Kutta method using an 'acceptable' formula if $\rho h < 1$ where $\rho = \max(1/\delta, \max_r |A_{rr} L|)$.

This is a weak result; the proof is immediate if the formula is explicit, and follows by considering a fixed-point iteration if the formula is semi-explicit.

Definition 3.1 A method of the form (2.6), with a parameter h, for computing values $\tilde{f}_j$ approximating $f(y_j)$ with $y_j \in [0, X]$ is said to be *convergent* on $[0, X)$ if, with the conventions of (2.4),

$$\lim_{h\to 0} \sup_{\substack{i\leq N-1 \\ Nh=X}} |f(y_j) - \tilde{f}_j| = 0. \tag{3.2}$$

Theorem 3.1 Subject to Condition 3.1: (a) an extended Runge-Kutta method (2.7) using an 'acceptable' Runge-Kutta formula is convergent, on $[0, X)$; (b) a mixed Runge-Kutta method using weights defined as in eqn (2.9) is convergent on $[0, X)$ if

$$\lim_{\substack{h\to 0 \\ Nh=X}} \{ \sup_{\substack{0\leq i\leq N-1 \\ x-ih \leq h}} | \int_0^{ih} F(x, y; f(y))dy - h \sum_{k=0}^{i} w_{i,k} F(x, kh; f(kh)) | \} = 0. \tag{3.3}$$

Condition 3.1, with (3.3), ensures the convergence of the quadrature method based on eqn (1.4). The proof of Theorem 3.1 follows similar lines to those given by Linz [14] - see also Baker [1] - to establish the convergence of (1.4). We require that with the chosen $\Omega_{j,k}$ in (2.6), and the convention (2.4),

$$\lim_{h\to 0} \sup_{0\leq i\leq N-1} |t_j| = 0, \tag{3.4}$$

where

$$t_j = \int_0^{y_j} F(y_j, y; f(y))dy - h \sum_{k=0}^{j} \Omega_{j,k} F(y_j, y_k; f(y_k)); \tag{3.5}$$

this is easily established.

Indeed, Theorem 3.1 is a corollary of the following result, in which we again assume Condition 3.1. Given $y_j = ih + \theta_r h$ with i, j, and r related by the conventions of (2.4), suppose that $0 \leq \theta_r \leq 1$ for $r = 0, 1, \ldots, p$, and $Nh = X$. Suppose further that $\sup_{0\leq k\leq j; i\leq N-1} |\Omega_{j,k}| = W < \infty$ and $\sup_{0\leq i\leq N-1} |t_j| = \tau$. Then for $i = 0, 1, \ldots, N-1$,

$$|\tilde{f}_j - f(y_j)| \leq \frac{\tau}{1-hLW} \exp\left(\frac{LWhj}{1-hLW}\right), \tag{3.6}$$

provided $hLW < 1$. Whilst (3.6) can be used to provide a strengthening of Theorem 3.1 - to show that the convergence in Theorem 3.1(a) is at least $O(h)$ for example - it is pessimistic in two respects:

(i) the rate of convergence predicted is pessimistic for extended high-order Runge-Kutta schemes and for high-order mixed formulae, where $r=p$, and (ii) for $j = 0, 1, 2, \ldots$ in sequence the error bound is exponentially increasing. We shall return to the stability aspect later, but note that such pessimism arises, in past, from the generality of the result.

It will be recalled that establishing the order of Runge-Kutta formulae for (1.5) is in general an involved procedure, and we may not expect any simplification when treating (1.1). Fortunately, the work of Pouzet [17] establishes† that the use of a formula of order q in the treatment of (1.5) gives the same high order convergence *at the points* $\{(ih + \theta_p h)\}$ in an extended method or with a suitable high-order mixed method, given sufficient smoothness. (Garey [8] effectively shows that considering a single first step is the critical stage in the argument.)

4. Stability

As mentioned in §1, stability of step-by-step quadrature methods for the numerical solution of (1.1) has been discussed by Noble [16] in terms of the work of Kobayasi [11] (summarized by Baker [1]). The class of methods considered by Kobayasi consists of those schemes (1.4) in which the weights $\{w_{i,k}\}$ are generated by repetition of a $(n + 1)$-point quadrature formula with variable modifications at the left and right ends (near $k = 0$, $k = i$ respectively), to take account of the fact that i is not always an integer multiple of n.

We suppose, with Kobayasi, that the approximations

$$\int_0^{ih} \phi(y)dy \simeq h \sum_{k=0}^{i} w_{i,k}\, \phi(kh) \qquad (i \geq i_0) \tag{4.1}$$

of eqn (1.2) are derived by constructing, for $\rho \equiv i \bmod(n)$, approximations

$$\int_0^{k_\rho h} \phi(y)dy \simeq h \sum_{k=0}^{k^*_\rho} \alpha_{\rho,k}\phi(kh), \qquad |\alpha_{\rho,k^*_\rho}| \neq 0, \tag{4.2}$$

$$\int_{k_\rho h+snh}^{k_\rho h+(s+1)nh} \phi(y)dy \simeq h \sum_{k=0}^{n} \gamma_k\, \phi((ns + k_\rho + k)h), \tag{4.3}$$

$$(s = 0, 1, \ldots, M_\rho -1;\ M_\rho = (i-\rho)/k)$$

$$\int_{(i-\rho+k_\rho)h}^{ih} \phi(y)dy \simeq h \sum_{k=0}^{k^{**}_\rho} \beta_{\rho,k}\phi((i - k)h), \ |\beta_{\rho,k^{**}_\rho}| \neq 0 \tag{4.4}$$

and summing these contributions. Suitable constraints are imposed on k_ρ, k^*_ρ and k^{**}_ρ, and $\Sigma\gamma_k = n$.

† The analysis of Pouzet for $p \leq 4$ can, apparently, be extended to general p.

Summarized briefly (and neglecting the effect of starting errors when $i_0 \neq 0$) the work of Kobayasi establishes that for suitable integers q, ν (with $\nu > q$) and appropriate functions $\psi_\rho(x)$

$$\tilde{f}(ih) = f(ih) + h^q \sum_{k=0}^{n-1} \xi_k^i e_k(ih) + O(h^\nu) \tag{4.5}$$

where

$$\xi_k = \exp(2\pi ki/n), \quad \lambda_t = \sum_{k=0}^{n} \gamma_k \xi_t^k/n, \quad K(x,y) = \frac{\partial}{\partial v} F(x,y;v)\Big]_{v=f(y)} \tag{4.6}$$

and

$$\sum_{k=0}^{n-1} \xi_k^\rho e_k(x) = \sum_{k=0}^{n-1} \lambda_k \xi_k^{k_\rho} \int_0^x K(x, y) e_k(y)dy - \psi_\rho(x) \tag{4.7}$$

$$(\rho = 0, 1, \ldots, n-1).$$

Here, k_ρ is associated with (4.2). (The presence of the functions $e_k(x)$ in (4.5) is due to the effect of truncation error and Kobayasi shows that starting errors produce a similar effect.)

An interpretation of (4.5) is provided by Noble [16] who (inter alia) shows that if $\underset{\sim}{G}^{-1} \underset{\sim}{H}$ is diagonalisable where $G_{\mu,\nu} = \xi_\nu^\mu$ and $H_{\mu,\nu} = \lambda_\nu \xi_\nu^{k_\mu}$ then the eigenvalues $\{\sigma_k\}$ of $\underset{\sim}{G}^{-1} \underset{\sim}{H}$ assume some importance in the stability theory. Thus we regard the algorithm as stable, when $F(x, y; v) = \lambda v$, if the eigenvalues satisfy $\sigma_0 = 1$ and $\lambda\sigma_k \lesssim \text{sign}(\lambda)$ for $k = 1, 2, \ldots, n-1$. *If the rules* (1.2) *have repetition factor one then* $\sigma_0 = 1$ *and* $\sigma_k = 0$ for $k = 1, \ldots, n-1$.

Remark. A perturbation $\varepsilon\gamma(x)$ in the function $g(x)$ of eqn (1.1) causes a change $\varepsilon e(x) + O(\varepsilon^2)$ in $f(x)$ where $e(x) = \int_0^x K(x, y) e(y)dy + \gamma(x)$. The use of (4.1) is shown to introduce a set of functions $e_k(x)$ associated with n parameters related to $\sigma_0, \sigma_1, \ldots, \sigma_{n-1}$ and some of these may produce unwanted parasitic solutions. The question of which components are undesirable can be decided simply only in the case $F(x, y; v) = \lambda v$ and $g(x) = 1$, but the 'optimum' situation is to ensure that $\sigma_k = 0$ for $k = 1, \ldots, n-1$.

The foregoing results suggest that it is an advantage to employ a quadrature method with repetition factor unity, and there are certainly examples (Baker [1]) of methods of repetition factor two in which the error growth is quite unacceptable.

Our summary of the work of Kobayasi and of Noble is motivated by the following observation: the analysis is immediately applicable to the extended or mixed methods when an equally-spacing Runge-Kutta formula is employed, and a completely analogous theory holds when using more general Runge-Kutta formulae. (The weights $\{w_{\mu,\nu}\}$ used in the mixed formulae are assumed to be generated in a similar way to those of (4.1).) The following result now assumes some significance.

Theorem 4.1. The weights $\{\Omega_{j,k}\}$ defined by (2.7) have repetition factor unity; those defined by (2.9) have repetition factor unity when the weights $\{w_{\mu,\nu}\}$ have repetition factor unity.

Not all schemes generated by the use of Runge-Kutta formulae have repetition factor one, and Baker and Keech [3] analyse a published algorithm with repetition factor twenty, and show it to be stable, in their sense, only if $h = 0$. As in their general study (expounded by Baker and Keech [2, 4]), attention is restricted to the equation

$$f(x) - \lambda \int_0^x f(y)dy = g(x) \qquad x \geqslant 0 \tag{4.8}$$

and a method is said to be *stable*† if successive values of the approximate solution can be grouped into vectors satisfying a relation $\underset{\sim}{\phi}_{k+1} = \underset{\sim}{M}\underset{\sim}{\phi}_k + \underset{\sim}{\gamma}_{k+1}$ where $\rho(\underset{\sim}{M})$, the spectral radius of $\underset{\sim}{M}$, satisfies the relation $\rho(\underset{\sim}{M}) \leqslant 1$ (with the restriction that $\underset{\sim}{M}$ is an M-matrix if $\rho(\underset{\sim}{M}) = 1$); the stability is said to be *strict* if $\rho(\underset{\sim}{M}) < 1$. A *region of stability* (respectively, strict stability) is a range of values of λh for which the given method is stable (respectively, strictly stable) if applied to (4.8).

<u>Theorem 4.2</u> The region of stability (strict stability) for an *extended* Runge-Kutta method for (4.8) is precisely the region of stability (strict stability) for the associated Runge-Kutta formula applied to $f'(x) = \lambda f(x)$, $x \geqslant 0$, with prescribed $f(0)$.

<u>Remark</u>. When comparing intervals of stability of extended Runge-Kutta methods with those of equal-step quadrature methods, the 'effective step-size' of the Runge-Kutta method can be borne in mind. Thus the Lawson form of fifth-order formula [12] is devized to have extended range of stability $(-5.5 \leqslant \lambda h \leqslant 0)$, but its parameters $\{\theta_r \mid r = 0, 1, \ldots, 6\}$ are $\{0, \frac{1}{2}, \frac{1}{4}, \frac{1}{2}, \frac{3}{4}, 1, 1\}$ so that its effective stepsize compared with an explicit quadrature method may be said to be $\frac{1}{4}h$ (or even $\frac{1}{6}h$, depending on interpretation).

As in the study of eqn (1.5) we can define *A-stability*. The scheme (2.6) is said to be A-stable if it is strictly stable - when applied to (4.8) - whenever $\mathrm{Re}(\lambda h) < 0$. Thus the extended Runge-Kutta method in Example 2.1(b) is A-stable. However, a mixed method (2.9) using the same Runge-Kutta formula need not be A-stable:

<u>Theorem 4.3</u> A mixed Runge-Kutta method (2.9) based on A-stable Runge-Kutta parameters is not necessarily A-stable.

A likely candidate to establish Theorem 4.3 is provided when we select a set of weights $\{w_{i,k}\}$ which have bad stability properties for use in eqn (2.9). Taking a combination of the trapezium rule and the repeated Simpson's rule (eqn (1.15) of [4]) with the formula in Example 2.1(b) provides a formula which is not A-stable. Keech (private communication) conjectures that the 'stable' combination of trapezium rule and Simpson's rule also destroys A-stability.

---

† Baker [1] also gives a definition of relative stability.

5. Implicit Runge-Kutta methods

The foregoing discussion has been limited somewhat artificially to the use of semi-explicit Runge-Kutta formulae. It is appropriate to observe that further generalisation is possible if we permit the use of implicit formulae. Such formulae, defined by the full tableau (2.1), are associated with quadrature rules

$$\int_0^{\theta_r h} \phi(y)dy \simeq h \sum_{t=0}^{p} A_{r,t}\, \phi(\theta_t h) \qquad (r = 0, 1, \ldots, p) \tag{5.1}$$

analogous to (2.3), and their use to discretize eqn (1.1) leads to formulae of the type

$$\tilde{f}_j - h \sum_k \Omega_{j,k}\, F(y_j, y_k; \tilde{f}_k) = g(y_j) \quad (j = 1, 2, \ldots) \tag{5.2}$$

where the summation over k now runs for $k = 0, 1, \ldots, (i+1)\times(p+1)$. Eqns (2.7) and (2.9) now define extended and mixed implicit Runge-Kutta methods when the suffices are permitted to vary as required, over $r, s = 0, 1, \ldots, p$.

Whilst Lemma 3.1 must be modified when treating implicit formulae derived as above, Theorems 3.1, 4.2, and 4.3 hold in the more general case, assuming $0 \leqslant \theta_r \leqslant 1$, $r = 0, 1, \ldots, p$.

Remark. Weiss [18] defines two classes of "implicit Runge-Kutta methods" without appealing directly to a Runge-Kutta tableau. The approach of Weiss is to generalize certain block-by-block methods of Linz and is based on the construction of interpolatory quadrature rules (5.1); the first class of methods forms (a subset of) the extended implicit Runge-Kutta methods as defined here. Whilst the use of equally-spaced values θ_r gives a method of Linz, A-stable methods can also be obtained by a suitable choice of $\theta_0, \theta_1, \ldots, \theta_p$ and here the work is related to techniques of Axelsson for treating (1.5). The second class of methods defined by Weiss (eqn (2.12) of [18]) overcomes - in a manner also adopted by Linz - an objection associated with (5.2). A difficulty arises when $F(x, y; v)$ is not available or is badly behaved for $y > x$, and is avoided by the use of a modified formula based on interpolation. A similar difficulty arises with semi-explicit formulae which are not 'convenient'.

6. Practical Aspects

The theoretical aspects outlined above dispose of the basic niceties of the mathematical analysis, and permit us to turn our attention to practical aspects. However, the practical testing of adaptive algorithms for integral equations remains in its infancy at the time of writing, and final judgement must await further practical work. We incline to the view, however, that Runge-Kutta methods may have been undervalued, in the past, by some writers.

Let us enumerate some questions which are not yet completely answered and are due to receive further attention:

(i) To what extent are existing stability theories adequate, and (given that A-stable methods are implicit or semi-explicit and hence expensive) when are A-stable methods desirable? (Whilst Weiss suggests that there are circumstances in which his implicit Runge-Kutta methods prove superior to step-by-step quadrature methods, our prejudices lead us to prefer semi-explicit formulae when suitable ones can be found.)

(ii) Can errors be controlled by varying the step-size or the order of formulae? (Care must be taken to avoid instability when making such changes.)
(iii) Can error estimates be computed satisfactorily (by using Fehlberg formulae [15], for example).
(iv) Is the development of new types of Runge-Kutta formulae, especially suited to (1.1), warranted?

Considering such questions, it would appear that whilst theoretical results have their value, the practical insight obtained from the present theory is somewhat limited. (Thus we have seen that convergent methods are not necessarily stable.) It should also be noted that the available stability theories are either concerned with a simple equation or are asymptotic in nature and take little account of a number of factors including the order (in h) of contributions from possibly unstable components. Finally, let us observe that in the numerical work quoted in [15], the given example of the Fehlberg formula appears to yield unsatisfactory error growth; more study is needed.

Example 6.1 We conclude with an example which may titillate the interest of the reader. In this we consider the equation

$$f(x) = ((1+x)\exp(-10x) + 1)^{\frac{1}{2}} + (1+x)(1-\exp(-10x) + 10\,\ell n\,(1+x)) - 10\int_0^x [(1+x)\{f(y)\}^2/(1+y)]dy,\ .$$

This equation was also considered by Weiss in [18], to illustrate the advantageous stability properties of his implicit methods based on Radau quadrature points. We tabulated errors obtained using a constant stepsize $h = 0.1$ with two extended Runge-Kutta methods:
(i) the semi-explicit method of Example 2.1(b) and (ii) an associated explicit formula defined by the tableau indicated: where θ_1' and θ_2' have the values given, in Example 2.1(b), to θ_0 and θ_1 respectively. The solution $f(x)$ has the form $f(x) = ((1+x)\exp(-10x) + 1)^{\frac{1}{2}}$.

$$\begin{array}{c|cccc} 0 & 0 & & & \\ \theta_1' & \theta_1' & 0 & & \\ \theta_2' & -\theta_2' & 2\theta_2' & 0 & \\ \hline 1 & 0 & \frac{1}{2} & \frac{1}{2} & 0 \end{array},$$

Errors in the first block are shown here:

	EXPLICIT	SEMI-EXPLICIT
$x = \theta_0 h$	-2.9×10^{-1}	7.8×10^{-2}
θ_1	5.0×10^{-1}	-3.6×10^{-2}
h	-5.3×10^{-1}	-4.8×10^{-2}

Errors at end-points of blocks are shown in the following figures. Only selected results are given, to economize space. (Numbers were obtained using a programme prepared by Miss Ruth Thomas.)

	EXPLICIT	SEMI-EXPLICIT
x = 5h	-1.3×10^{-2}	-5.2×10^{-3}
10h	-1.0×10^{-2}	-3.0×10^{-3}
15h	-8.1×10^{-3}	-2.4×10^{-3}
20h	-6.7×10^{-3}	-2.0×10^{-3}
25h	-5.7×10^{-3}	-1.7×10^{-3}
50h	-3.3×10^{-3}	-9.7×10^{-4}
100h	-1.8×10^{-3}	-5.2×10^{-4}

Table of Errors, Constant stepsize h = 0.1

It does not appear to be possible to obtain good accuracy over initial steps with large h, and the only way to exploit a good stability interval seems to involve variable h.

7. References

[1] Baker, C.T.H. The numerical treatment of integral equations. Clarendon Press, Oxford (in press).

[2] Baker, C.T.H. and Keech, M.S. Regions of stability in the numerical treatment of Volterra integral equations. Numerical Analysis Report No.12, Department of Mathematics, University of Manchester (1975).

[3] Baker, C.T.H., and Keech, M.S. On the instability of a certain Runge-Kutta procedure for a Volterra integral equation. Numerical Analysis Report No.21, Department of Mathematics, University of Manchester (1977).

[4] Baker, C.T.H. and Keech, M.S. Stability regions in the numerical treatment of Volterra integral equations. SIAM J. Numer. Anal. (to appear).

[5] Beltyukov, B.A. An analogue of the Runge-Kutta methods for the solution of a non-linear equation of the Volterra type. (Translation:) Differential Equations 1 pp. 417-433 (1965).

[6] Butcher, J.C. Implicit Runge-Kutta processes. Math. Comp. 18 pp. 50-64 (1964).

[7] Delves, L.M. and Walsh, J.E. (editors) Numerical solution of integral equations. Clarendon Press, Oxford (1974).

[8] Garey, L. Solving nonlinear second kind Volterra equations by modified increment methods. SIAM J. Numer. Anal. 12 pp.501-508 (1975).

[9] Hall, G. and Watt, J.M. (editors). Modern numerical methods for ordinary differential equations. Clarendon Press, Oxford (1976).

[10] Henrici, P. Discrete variable methods in ordinary differential equations. Wiley, New York (1962).

[11] Kobayasi, M. On the numerical solution of the Volterra integral equations of the second kind by linear multistep methods. Rep. Stat. Appl. Res. JUSE. 13 pp. 1-21 (1966).

[12] Lambert, J.D. Computational methods in ordinary differential equations. Wiley, New York (1973).

[13] Lapidus, L. and Seinfeld, J.H. Numerical solution of ordinary differential equations. Academic Press, New York, (1971).

[14] Linz, P. The numerical solution of Volterra integral equations by finite difference methods. MRC Tech. Summary Report No.825, Madison, Wisc. (1967).

[15] Lomakovič, A.M. and Iščuk, V.A. An approximate solution of a non-linear integral equation of Volterra type by a two sided Runge-Kutta-Fehlberg method. Vyč. i Prik. Mat. 23 pp.29-40 (1974). In Russian.

[16] Noble, B. Instability when solving Volterra integral equations of the second kind by multistep methods. Lecture Notes in Mathematics 109, Springer-Verlag, Berlin (1969).

[17] Pouzet, P., Étude, en vue de leur traitment numérique des équations integrales de type Volterra. Revue Francaise de traitment de l'information, 6 pp.79-112 (1963).

[18] Weiss, R. Numerical procedures for Volterra integral equations Ph.D. thesis, Australian National University, Canberra.

BEST APPROXIMATION OF COMPLEX-VALUED DATA

Ian Barrodale

Abstract

We consider problems arising in the determination of best approximations to complex-valued data. The emphasis is on linear approximation in the ℓ_1 and ℓ_∞ norms, but some remarks on ℓ_∞ rational approximation are also included.

1. Introduction

The general complex linear discrete approximation problem can be stated as follows. Let $f(z)$ and $\phi_1(z), \phi_2(z), \ldots, \phi_n(z)$ be given complex-valued functions defined on a discrete subset $Z = \{z_t | t = 1,2,\ldots,m \text{ , } m \geq n\}$ of complex N-dimensional space C^N. Also, for any set $A = \{a_1, a_2, \ldots, a_n\}$ of complex parameters, let $L(A,z) = \sum_{j=1}^{n} a_j\phi_j(z)$ and $r(A,z) = f(z) - L(A,z)$. Then, for a given norm $||\cdot||$, the problem is to determine a best parameter set $\hat{A}$ for which

$$(1.1) \qquad ||r(\hat{A},z)|| \leq ||r(A,z)|| \text{ , } \qquad \text{for each } A \text{ .}$$

An $\hat{A}$ satisfying (1.1) always exists (e.g. see Meinardus (1967)), but in general it may not be unique. $L(\hat{A},z)$ is called a best approximation with respect to the given norm.

In the real case (i.e. real-valued functions with real parameters) the theory of best approximation in the ℓ_1, ℓ_2, and ℓ_∞ norms is well developed, and reliable algorithms are available to determine best approximations in all three norms. In the complex case the ℓ_1, ℓ_2, and ℓ_∞ norms are defined, respectively, as follows:

$$(1.2) \qquad ||g(z)||_1 = \sum_{t=1}^{m} |g(z_t)| \text{ ,}$$

$$(1.3) \qquad ||g(z)||_2^2 = \sum_{t=1}^{m} |g(z_t)|^2 \text{ ,}$$

$$(1.4) \qquad ||g(z)||_\infty = \max_{1 \leq t \leq m} |g(z_t)| \text{ .}$$

Here, $g(z)$ is a complex-valued function defined on Z and $|\cdot|$ is a modulus sign.

Replacing the real ℓ_2 norm by its complex analogue (1.3) is quite straightforward. In particular, best complex least-squares approximations can be computed either by employing complex arithmetic versions of the algorithms for determining real ℓ_2 approximations, or by restating the complex problem, via its real and imaginary parts, as an equivalent real problem of larger dimensions.

We shall make no further reference in this paper to ℓ_2 approximation.

2. Linear ℓ_1 approximation

The complex ℓ_1 problem is to determine a parameter set A_1 to minimize

(2.1) $$||r(A,z)||_1 = \sum_{t=1}^{m} |f(z_t) - L(A,z_t)| .$$

It is easy to prove that $||r(A,z)||_1$ is a convex function of A, although it is not differentiable if $f(z_t) = L(A,z_t)$ for some value of t.

In the real case the best ℓ_1 approximation (or one of them, in the event of nonuniqueness) interpolates f on Z at least k times, where k is the rank of the $m \times n$ matrix $[\phi_j(z_t)]^T$: see Barrodale and Roberts (1970). Stuart (1973) provides the following example which shows that a best complex ℓ_1 approximation does not necessarily interpolate the given function f at any point of Z.

Example 1. Compute the best ℓ_1 approximation by $L(A,z) = a_1 + a_2 z$ to $f(z) = z^2$ on $Z = \{z|\ \mathrm{Re}(z) = 0(1)2\ ,\ \mathrm{Im}(z) = 0(.5)1\}$.

The smallest ℓ_1 error that can be obtained by interpolating on all possible pairs of points of Z is 7.623, whereas an ℓ_1 error of 7.123 occurs when $a_1 = -.2582 - i$ and $a_2 = 2 + i$. These parameter values are optimal, and they yield a unique best ℓ_1 approximation which interpolates f at no point of Z.

Computational experience in minimizing (2.1) indicates that a best complex ℓ_1 approximation may or may not be interpolatory, and hence no simple characteristic property appears to be available. Also, it is our experience that expression (2.1) is quite difficult to minimize in practice. We are thus led to consider a simpler alternative problem.

Let v be a vector with m complex components $v_t = x_t + iy_t$, and let us define a norm $||\cdot||_\dagger$ as

(2.2) $$||v||_\dagger = \sum_{t=1}^{m} (|x_t| + |y_t|) .$$

The complex ℓ_1 norm can be estimated by the norm (2.2), in view of the relationship

(2.3) $$||v||_1 \le ||v||_\dagger \le \sqrt{2}\ ||v||_1 .$$

The corresponding best approximation problem is thus to determine $A_\dagger$, for which

(2.4) $$||r(A_\dagger,z)||_\dagger \le ||r(A,z)||_\dagger , \qquad \text{for each } A .$$

As is explained below, $A_\dagger$ can be easily obtained as the solution to a linear problem, so we should examine the consequences of substituting $A_\dagger$ in place of A_1 in the ℓ_1 problem.

<u>Theorem 1</u>: $||r(A_1,z)||_1 \le ||r(A_\dagger,z)||_1 \le \sqrt{2}\,||r(A_1,z)||_1$.

Proof.
$$\begin{aligned} ||r(A_1,z)||_1 &\le ||r(A_\dagger,z)||_1 && \text{from (2.1)} \\ &\le ||r(A_\dagger,z)||_\dagger && \text{from (2.3)} \\ &\le ||r(A_1,z)||_\dagger && \text{from (2.4)} \\ &\le \sqrt{2}\,||r(A_1,z)||_1 && \text{from (2.3)} \end{aligned}$$

<u>Theorem 2</u>: $\frac{1}{\sqrt{2}}||r(A_\dagger,z)||_\dagger \le ||r(A_1,z)||_1 \le ||r(A_\dagger,z)||_\dagger$

Proof.
$$\begin{aligned} \frac{1}{\sqrt{2}}||r(A_\dagger,z)||_\dagger &\le \frac{1}{\sqrt{2}}||r(A_1,z)||_\dagger && \text{from (2.4)} \\ &\le ||r(A_1,z)||_1 && \text{from (2.3)} \\ &\le ||r(A_\dagger,z)||_1 && \text{from (2.1)} \\ &\le ||r(A_\dagger,z)||_\dagger && \text{from (2.3)} \end{aligned}$$

Thus, $L(A_\dagger,z)$ is a near-best approximation for which the ℓ_1 error is within a factor of $\sqrt{2}$ of its minimum value $||r(A_1,z)||_1$, which in turn can be well estimated by $||r(A_\dagger,z)||_\dagger$. It therefore seems likely that in any practical application it is sufficient to solve problem (2.4), rather than the more difficult problem (2.1).

An $A_\dagger$ satisfying (2.4) can be obtained by minimizing

(2.5) $$\sum_{t=1}^{m}\left[\left|\mathrm{Re}\big(f(z_t) - L(A,z_t)\big)\right| + \left|\mathrm{Im}\big(f(z_t) - L(A,z_t)\big)\right|\right]$$

by linear programming. However, expression (2.5) can also be minimized using any algorithm (of linear programming type, or otherwise) which solves an overdetermined system of M real linear equations in N real unknowns, say

(2.6) $$\underset{\sim}{B}\,\underset{\sim}{x} = \underset{\sim}{g} \ ,$$

in the (real) ℓ_1 sense. The details are as follows.

Letting

(2.7) $$a_j = b_j + ic_j \ ,$$

(2.8) $$f(z_t) = d(z_t) + ie(z_t) = d_t + ie_t \ ,$$

(2.9) $$\phi_j(z_t) = h_j(z_t) + ik_j(z_t) = h_{j,t} + ik_{j,t} \ ,$$

and putting $R(A,z_t) = \mathrm{Re}\big(r(A,z_t)\big)$ and $I(A,z_t) = \mathrm{Im}\big(r(A,z_t)\big)$, we have

(2.10) $$R(A,z_t) = d_t - \sum_{j=1}^{n}\big(b_jh_{j,t} - c_jk_{j,t}\big) \ , \quad \text{and}$$

$$(2.11) \qquad I(A,z_t) = e_t - \sum_{j=1}^{n} \left(b_j k_{j,t} + c_j h_{j,t}\right) .$$

Then, defining two $m \times n$ matrices

$$(2.12) \qquad \underset{\sim}{H} = [h_{j,t}]^T \text{ and } \underset{\sim}{K} = [k_{j,t}]^T , \text{ and two vectors}$$

$$(2.13) \qquad \underset{\sim}{d} = [d_1,d_2,\ldots,d_m]^T \text{ and } \underset{\sim}{e} = [e_1,e_2,\ldots,e_m]^T ,$$

problem (2.4) can be solved in the form (2.6) by setting

$$(2.14) \qquad \underset{\sim}{B} = \left[\begin{array}{c|c} \underset{\sim}{H} & -\underset{\sim}{K} \\ \hline \underset{\sim}{K} & \underset{\sim}{H} \end{array}\right] \text{ and } \underset{\sim}{g} = \left[\begin{array}{c} \underset{\sim}{d} \\ \hline \underset{\sim}{e} \end{array}\right] .$$

Here, $M = 2m$ and $N = 2n$, and an ℓ_1 solution $\underset{\sim}{x}^\dagger$ to (2.6) has the form

$$(2.15) \qquad \underset{\sim}{x}^\dagger = \left[b_1^\dagger,b_2^\dagger,\ldots,b_n^\dagger,c_1^\dagger,c_2^\dagger,\ldots,c_n^\dagger\right]^T .$$

$A_\dagger$ is formed from (2.15) using (2.7). In the event that the parameters a_j are required to be real, then $N = n$ and $\underset{\sim}{B}$ is set equal to the left half of the partitioned matrix in (2.14).

By posing problem (2.4) in the form (2.6) we see that $L(A_\dagger,z)$ must satisfy the following interpolatory property; this is an immediate consequence of the interpolatory property referred to in the second paragraph of this section.

Theorem 3: Let the rank of $\underset{\sim}{B}$ be ρ (and note that, normally, $\rho = 2n$ for complex-valued a_j's, and $\rho = n$ for real-valued a_j's). Then at least ρ of the $2m$ quantities

$$|R(A_\dagger,z_t)| \text{ and } |I(A_\dagger,z_t)|$$

are zero, for some best approximation $L(A_\dagger,z)$.

In summary, in view of Theorems 1 and 2 and the computational effort required to minimize expression (2.1), we are convinced that for all practical purposes problem (2.6) should be solved instead of the complex ℓ_1 problem. We have solved several test problems of type (2.4) by applying the Fortran program given by Barrodale and Roberts (1974) to the equivalent real problems of type (2.6).

3. Linear ℓ_∞ approximation

The complex ℓ_∞ problem is to determine a parameter set A_∞ which minimizes

$$(3.1) \qquad ||r(A,z)||_\infty = \max_{1 \le t \le m} |r(A,z_t)| .$$

In contrast with the complex ℓ_1 problem, problem (3.1) has been considered by several authors. For example, Meinardus (1967) provides proofs of the following two theorems.

Theorem 4 (Kolmogoroff): $L(A_\infty,z)$ is a best ℓ_∞ approximation if and only if the inequality

$$\min_{z\in Z_0} \mathrm{Re}\left\{\overline{\left(f(z) - L(A\ ,z)\right)}\ L(A,z)\right\} \le 0$$

holds for every A, where Z_0 is an extremal set defined by

$$Z_0 = \left\{z\in Z \,\middle|\, |f(z) - L(A_\infty,z)| = ||f(z) - L(A_\infty,z)||_\infty\right\} .$$

Theorem 5 (Haar; Kolmogoroff): If $\{\phi_1,\phi_2,\ldots,\phi_n\}$ satisfies the Haar condition on Z (i.e. if every $L(A,z)$ which is not identically zero vanishes at no more than $n - 1$ points of Z), then for any f defined on Z there exists precisely one best ℓ_∞ approximation.

In the real case Theorem 4 can be used to show that, when the Haar condition holds, the best ℓ_∞ approximation determines (and can be determined by) an extremal set containing exactly $n + 1$ points. However, in the complex case the number of points p in the extremal set satisfies the inequality $n + 1 \le p \le 2n + 1$, i.e. the exact value of p is unknown a priori. Williams (1972) presents an exchange-type complex algorithm which assumes that $p = n+1$, but in general his method does not yield best approximations.

Ellacott and Williams (1976a) use a complex version of Lawson's algorithm to solve problem (3.1), and although their method is slowly convergent they state that they are unaware of an algorithm which is faster. Finally, noting that an A_∞ can be determined by solving the nonlinear convex programming problem

$$\text{(3.2)} \qquad \min_{A,w}\{w \mid [R(A,z_t)]^2 + [I(A,z_t)]^2 \le w\ , \quad \text{for} \quad t = 1,2,\ldots,m\}\ ,$$

Barrodale, Delves and Mason (1977) present an algorithm for solving problem (3.1) which is sometimes rapidly convergent. This algorithm consists of replacing each quadratic constraint in (3.2) by its linear Taylor approximation, solving the resulting linear programming problem, and repeating the procedure until convergence occurs. However, for some test problems, this method is also slowly convergent. Hopefully this deficiency can be rectified by modifying the algorithm so that it takes some account of the constraint curvature in problem (3.2).

It is apparent that the nonlinearity in (3.1) and (3.2) is due to the choice of norm, so again we are led to consider a simpler related problem. For any vector v with m complex components $v_t = x_t + iy_t$, let $||\cdot||_*$ be a norm defined by

$$\text{(3.3)} \qquad ||v||_* = \max_{1\le t\le m} \{\max(|x_t|,|y_t|)\} .$$

Observing that

(3.4) $$||v||_* \le ||v||_\infty \le \sqrt{2}\,||v||_* ,$$

Barrodale, Delves and Mason (1977) provide the three theorems below relating problem (3.1) to the problem of determining a parameter set A_* for which

(3.5) $$||r(A_*,z)||_* \le ||r(A,z)||_* , \quad \text{for each } A .$$

Theorem 6: $$||r(A_\infty,z)||_\infty \le ||r(A_*,z)||_\infty \le \sqrt{2}\,||r(A_\infty,z)||_\infty .$$

Theorem 7: $$||r(A_*,z)||_* \le ||r(A_\infty,z)||_\infty \le \sqrt{2}\,||r(A_*,z)||_* .$$

An A_* satisfying (3.5) can be obtained by solving the linear programming problem

$$\min_{A,u}\left\{u \;\middle|\; |R(A,z_t)| \le u ,\; |I(A,z_t)| \le u , \quad \text{for } t = 1,2,\ldots,m\right\} .$$

However, again it is more convenient to solve the overdetermined system (2.6). Here $\underset{\sim}{B}$ and $\underset{\sim}{g}$ remain as defined in (2.14), but now we calculate the (real) ℓ_∞ solution $\underset{\sim}{x}^*$, which yields the following extremal property.

Theorem 8: Let the rank of $\underset{\sim}{B}$ be ρ. Then for some best approximation $L(A_*,z)$ the $2m$ inequalities

$$|R(A_*,z_t)| \le u , \quad |I(A_*,z_t)| \le u$$

become actual equalities at least $\rho + 1$ times when u assumes its minimum value $||r(A_*,z)||_*$.

Hence, rather than solving the nonlinear problem (3.1), in practice the linear problem (3.5) should probably be solved instead. We have used the Fortran program in Barrodale and Phillips (1975) for this purpose. Of course, if problem (3.1) has to be solved, the solution A_* to problem (3.5) provides a good initial estimate of A_∞ in view of Theorems 6 and 7.

4. A numerical example

In order to illustrate the use of Theorems 1 and 2 in ℓ_1 approximation, and Theorems 6 and 7 in ℓ_∞ approximation, we present the following numerical example.

Example 2. Let $f(z) = 1/[z - (2+i)]$,

$$Z = \left\{z_t \;\middle|\; z_t = e^{i\pi(t-1)/20} , \quad \text{for } t = 1,2,\ldots,40\right\} ,$$

$$\text{and } L(A,z) = \sum_{j=1}^{n} a_j z^{j-1} , \quad a_j \text{ complex-valued, for } n = 2,4,6,8.$$

Solve problem (2.1) to obtain $L(A_1,z)$, problem (2.4) to obtain $L(A_\dagger,z)$, problem (3.1) to obtain $L(A_\infty,z)$, and problem (3.4) to obtain $L(A_*,z)$, and give the values of

$$\ell_{\dagger\dagger} = \frac{1}{40}\,||r(A_\dagger,z)||_\dagger \qquad \ell_{**} = ||r(A_*,z)||_*$$

$$\ell_{\dagger 1} = \frac{1}{40}\,||r(A_{\dagger},z)||_1 \qquad \ell_{*\infty} = ||r(A_*,z)||_\infty$$

$$\ell_{11} = \frac{1}{40}\,||r(A_1,z)||_1 \qquad \ell_{\infty\infty} = ||r(A_\infty,z)||_\infty \ .$$

The results are as follows:

n = 2	$\ell_{\dagger\dagger}$ = .11431 (0)	ℓ_{**} = .11177 (0)
	$\ell_{\dagger 1}$ = .89844 (-1)	$\ell_{*\infty}$ = .11182 (0)
	ℓ_{11} = .89622 (-1)	$\ell_{\infty\infty}$ = .11180 (0)
n = 4	$\ell_{\dagger\dagger}$ = .22859 (-1)	ℓ_{**} = .21899 (-1)
	$\ell_{\dagger 1}$ = .18207 (-1)	$\ell_{*\infty}$ = .22936 (-1)
	ℓ_{11} = .17890 (-1)	$\ell_{\infty\infty}$ = .22361 (-1)
n = 6	$\ell_{\dagger\dagger}$ = .45668 (-2)	ℓ_{**} = .44023 (-2)
	$\ell_{\dagger 1}$ = .37012 (-2)	$\ell_{*\infty}$ = .47856 (-2)
	ℓ_{11} = .35777 (-2)	$\ell_{\infty\infty}$ = .44721 (-2)
n = 8	$\ell_{\dagger\dagger}$ = .87199 (-3)	ℓ_{**} = .85658 (-3)
	$\ell_{\dagger 1}$ = .73009 (-3)	$\ell_{*\infty}$ = .10078 (-2)
	ℓ_{11} = .71554 (-3)	$\ell_{\infty\infty}$ = .89443 (-3)

In this example the relevant inequalities in Theorems 1, 2, 6, and 7 hold to within a factor which is considerably less than $\sqrt{2}$, although in other numerical examples that we have encountered the factor $\sqrt{2}$ has been achieved.

5. Rational ℓ_∞ approximation

The complex rational discrete approximation problem is to determine a best approximation to $f(z)$ on Z by functions of the form

$$F(A,z) = \sum_{j=1}^{s} p_j\phi_j(z) \Big/ \sum_{k=1}^{n} q_k\psi_k(z) \ . \tag{5.1}$$

In (5.1) we denote by $A = \{p_1,p_2,\ldots,p_s,q_1,q_2,\ldots,q_n\}$ the set of complex parameters, and each ϕ_j and ψ_k is a given complex-valued function.

In the real case the theory and algorithms for best rational approximation have been developed mainly for the ℓ_∞ norm (see Barrodale and Roberts (1973) for a recent survey), so we shall refer only to complex ℓ_∞ approximation here.

Ellacott and Williams (1976b) discuss complex rational ℓ_∞ approximation, and we refer the interested reader to their paper for a more detailed exposition than is given here. They present a descent algorithm for computing local best approximations which employs a complex version of Lawson's algorithm, and, as the authors themselves note, the efficiency of their method could be improved through the use of a fast algorithm for complex linear ℓ_∞ approximation. The concept of a local best approximation is indeed necessary in the complex rational ℓ_∞ problem, as is apparent from the following remarks.

Letting $Z = \{-2, -1, -\frac{1}{2}, \frac{1}{2}, 1, 2\}$, and defining f on Z by the function values $f(-2) = f(2) = -\frac{1}{3}$, $f(-1) = f(1) = \frac{7}{9}$, and $f(-\frac{1}{2}) = f(\frac{1}{2}) = \frac{4}{3}$, then by using the complex rational approximating function $p_1/(q_1 + q_2 z)$, Powell (1977) has shown for this example that

(i) the best ℓ_∞ approximation has complex parameters (i.e. when the parameters p_1, q_1, and q_2 are restricted to be real, the corresponding (real) ℓ_∞ error is increased),

(ii) the best ℓ_∞ approximation is not unique, by symmetry,

(iii) the data can be perturbed to produce a local best approximation that is not a global best approximation.

Furthermore, these results remain valid if the ℓ_∞ norm is replaced by the norm $||\cdot||_*$ defined in (3.3). (Result (i) above, which states that allowing complex parameters in rational approximations to real data can reduce the error, may also have been discovered independently by others.)

Finally, we wish to report on an attempt to apply to the complex case the ideas of Roberts (1977) regarding computation of minimal degree approximations. His approach enables real rational approximations of specified accuracy to be efficiently obtained by solving a finite sequence of linear problems. For complex rational approximation an analogous technique requires the solution of a finite sequence of integer programming problems, an approach which can hardly be described as efficient.

Acknowledgements. The author is indebted to Dr. J.C. Mason for introducing him to this general problem and for collaboration thereafter, to Professor M.J.D. Powell for permission to use the private communication summarized in Section 5, and especially to Professor L.M. Delves with whom the author has collaborated extensively in connection with the material of this paper. The financial assistance provided by NRC Grant No. A5251 is also gratefully acknowledged.

REFERENCES

I. Barrodale, L.M. Delves and J.C. Mason (1977), Linear Chebyshev Approximation of Complex-Valued Functions, Math. Dept. Report No. 89, Univ. of Victoria, Victoria, B.C., Canada.

I. Barrodale and C. Phillips (1975), Algorithm 495 - Solution of an Overdetermined System of Linear Equations in the Chebyshev Norm, TOMS, v. 1, pp. 264-270.

I. Barrodale and F.D.K. Roberts (1970), Applications of Mathematical Programming to ℓ_p Approximation, from "Nonlinear Programming", edited by J.B. Rosen, O.L. Mangasarian, and K. Ritter. Academic Press, pp. 447-464.

I. Barrodale and F.D.K. Roberts (1973), Best Approximation by Rational Functions, from "Proceedings of the 3rd Manitoba Conference on Numerical Mathematics", Univ. of Manitoba, Winnipeg, Manitoba, Canada. pp. 3-29.

I. Barrodale and F.D.K. Roberts (1974), Algorithm 478 - Solution of an Overdetermined System of Equations in the ℓ_1 Norm, Comm. ACM, v. 17, pp. 319-320.

S. Ellacott and J. Williams (1976a), Linear Chebyshev Approximation in the Complex Plane Using Lawson's Algorithm, Math. Comp., v. 30, pp. 35-44.

S. Ellacott and J. Williams (1976b), Rational Chebyshev Approximation in the Complex Plane, SIAM J. Numer. Anal., v. 13, pp. 310-323.

G. Meinardus (1967), "Approximation of Functions: Theory and Numerical Methods", Springer-Verlag, New York.

M.J.D. Powell (1977) Private communication.

F.D.K. Roberts (1977), Minimal Degree Rational Approximation, Rocky Mountain J. of Math., to appear.

G.F. Stuart "Numerical ℓ_1 and ℓ_∞ Approximation", M.Sc. thesis, Univ. of Western Ontario, London, Ontario, Canada.

J. Williams (1972), Numerical Chebyshev Approximation in the Complex Plane, SIAM J. Numer. Anal., v. 9, pp. 638-649.

INVERSE EIGENVALUE PROBLEMS FOR BAND MATRICES

D. Boley and G. H. Golub*

0. Introduction

We are interested in solving the inverse eigenvalue problem associated with band matrices. We shall place particular emphasis on giving a stable numerical algorithm for determining such matrices.

Let A be a real, symmetric matrix with

$$a_{ij} = 0 \quad \text{when} \quad |i - j| \geq p \ .$$

Define

$$\{A^{(k)}\}_{i,j} = a_{ij} \qquad \begin{pmatrix} i = k,k+1,\dots,n \ ; \\ j = k,k+1,\dots,n \end{pmatrix} \ ,$$

so that $A^{(k)}$ is derived from A by striking out the first $(k-1)$ rows and columns of A. We denote the eigenvalues of $A^{(k)}$ as $\{\lambda_i^{(k)}\}_{i-1}^{n-k+1}$ and, we assume

$$\lambda_i^{(k)} < \lambda_{i+1}^{(k)} \qquad (i = 1,2,\dots,n-k+2)$$

and

$$\lambda_i^{(k)} < \lambda_i^{(k+1)} < \lambda_{i+1}^{(k)} \ .$$

The problem then is to determine the matrix A from the p sets of eigenvalues $\{\lambda_i^{(k)}\}_{i=1}^{n-k+1}$, $(k = 1,2,\dots,p)$. Note that the number of unknowns is precisely equal to the number of eigenvalues specified.

For A tri-diagonal $(p = 2)$, the problem has been studied extensively (cf. [1], [2], [3]). The numerical algorithm derived in this paper is similar to that given in [3] but the method for deriving the algorithm is different.

In order to simplify the notation and discussion, we shall primarily concern ourselves with five diagonal matrices in this paper and discuss the general case in

---

*The work of this author was supported in part by the NSF MCS75-13497-A01 and in part by ERDA EY-76-S-03-0326PA#30.

a future paper. We use the following notation. Let

$$K \equiv A = \begin{bmatrix} a_1 & b_1 & c_1 & & & & \\ b_1 & a_2 & b_2 & c_2 & & \bigcirc & \\ c_1 & b_2 & \cdot & \cdot & \cdot & & \\ & c_2 & \cdot & \cdot & \cdot & \cdot & \\ & & \cdot & \cdot & \cdot & \cdot & c_{n-2} \\ & \bigcirc & & \cdot & \cdot & \cdot & b_{n-1} \\ & & & & c_{n-2} & b_{n-1} & a_n \end{bmatrix},$$

$$L \equiv A^{(2)} = \begin{bmatrix} a_2 & b_2 & c_2 & & \bigcirc \\ b_2 & a_3 & b_3 & \cdot & \\ c_2 & b_3 & \cdot & \cdot & c_{n-2} \\ & \cdot & \cdot & \cdot & b_{n-1} \\ \bigcirc & & c_{n-2} & b_{n-1} & a_n \end{bmatrix}, \qquad M \equiv A^{(3)} = \begin{bmatrix} a_3 & b_3 & c_3 & & \bigcirc \\ b_3 & a_4 & & \cdot & \\ c_3 & \cdot & \cdot & & c_{n-2} \\ & \cdot & \cdot & \cdot & b_{n-1} \\ \bigcirc & & c_{n-2} & b_{n-1} & a_n \end{bmatrix};$$

we assume the integer n is even. Let the eigenvalues of K, L, M be $\{\kappa_i\}_{i=1}^{n}$, $\{\lambda_i\}_{i=1}^{n-1}$, $\{\mu_i\}_{i=1}^{n-2}$, respectively. We assume

$$\kappa_i < \kappa_{i+1}, \qquad \lambda_i < \lambda_{i+1}, \qquad \mu_i < \mu_{i+1}$$

and

$$\kappa_i < \lambda_i < \kappa_{i+1} \qquad\qquad \lambda_i < \mu_i < \lambda_{i+1} .$$

In section 1, we give the solution of a related problem which is required for the final solution. The block Lanczos algorithm is described in section 2 since it plays a fundamental role in generating K. Finally in section 3, we describe the method for generating $K = A$.

1. Solution of Related Problem

In order to determine K from its eigenvalues we need the solution to a related problem. The solution to this related problem is derived in [4] and we summarize it here.

Let A be a real symmetric matrix with distinct eigenvalues $\{\alpha_i\}_{i=1}^{n}$ and let Q be the matrix of eigenvectors.

Suppose we are given a set of values $\{\beta_i\}_{i=1}^{n-1}$ $(\beta_i < \beta_{i+1})$ with

(1.1) $$\alpha_i < \beta_i < \alpha_{i+1} .$$

We wish to determine the vector $\underset{\sim}{c}$ so that $\{\beta_i\}_{i=1}^{n-1}$ are the stationary values of

$$\underset{\sim}{x}^T A \underset{\sim}{x}$$

subject to

(1.2) $$\begin{cases} \underset{\sim}{x}^T \underset{\sim}{x} = 1 , \\ \underset{\sim}{c}^T \underset{\sim}{x} = 0 , \\ \underset{\sim}{c}^T \underset{\sim}{c} = 1 . \end{cases}$$

In [4], it is shown that if

$$\underset{\sim}{d} = Q^T \underset{\sim}{c}$$

then

(1.3) $$d_k^2 = \frac{\prod_{j=1}^{n-1} (\beta_j - \alpha_k)}{\prod_{\substack{j=1 \\ j \neq k}}^{n} (\alpha_j - \alpha_k)} .$$

Note that (1.1) guarantees that the right hand side of (1.2) is positive. We may assign d_k a positive or negative value so that there are 2^{n-1} solutions.

If the eigenvalues and stationary values are specified <u>and</u> $\underset{\sim}{c}$ is specified then we may deduce information about the matrix of eigenvectors. In particular if $\underset{\sim}{c} = \underset{\sim}{e}_1$, $e_1 = (1,0,\ldots,0)^T$, then the first row of the matrix of eigenvectors is determined; <u>viz.</u>

(1.4) $$q_{1j}^2 = \frac{\prod_{k=1}^{n-1} (\beta_k - \alpha_j)}{\prod_{\substack{k=1 \\ k \neq j}}^{n} (\alpha_k - \alpha_j)} .$$

If $\underset{\sim}{c} = \underset{\sim}{e}_1$, then the stationary values of $\underset{\sim}{x}^T A \underset{\sim}{x}$ subject to the constraints (1.2) are the eigenvalues of $A^{(2)}$ the matrix obtained from A by striking out its first row and column. This is true whether or not A is a band matrix. Thus we have the following:

Lemma. Let A be a real symmetric matrix with distinct eigenvalues $\{\lambda_i^{(1)}\}_{i=1}^n$, and $A^{(2)}$ the matrix $\{a_{ij}\}_{i,j=2}^n$ with eigenvalues $\{\lambda_i^{(2)}\}_{i=1}^{n-1}$. If $A = Q \Lambda Q^T$, then

(1.5) $$q_{1j}^2 = \frac{\prod_{k=1}^{n-1} (\lambda_k^{(2)} - \lambda_j^{(1)})}{\prod_{\substack{k=1 \\ k \neq j}}^{n} (\lambda_k^{(1)} - \lambda_j^{(1)})} .$$

2. The Block Lanczos Algorithm

The Lanczos algorithm (cf. [5]) is a very useful method for determining the eigenvalues of a real symmetric matrix. The basic idea is to generate from a given matrix A a tri-diagonal matrix J which has the same eigenvalues as A. Thus, given A, one determines the columns of the matrix X and a matrix J so that

$$A = X J X^T$$

where

$$X^T X = I$$

and

$$J \text{ is a Jacobi or tri-diagonal matrix } .$$

In some applications, it is desirable that the generated matrix J be block tri-diagonal. The calculation proceeds in the following way.

Block Lanczos Algorithm: Given r such that $r\ell = n$ and X_1 an $n \times r$ matrix such that $X_1^T X_1 = I_r$, compute sequence of matrices $X_2, X_3, \ldots, X_\ell$, $M_1, \ldots, M_\ell$ and $R_2, R_3, \ldots, R_\ell$.

Step 1. Compute AX_1 and $M_1 = X_1^T A X_1$.

Step 2. For $i = 1, 2, \ldots, \ell-1$

2.a Compute

$$Z_{i+1} = \begin{cases} AX_1 - X_1 M_1 & \text{for } i = 1 \\ AX_i - X_i M_i - X_{i-1} R_i^T & \text{for } i > 1 . \end{cases}$$

2.b Compute X_{i+1} and R_{i+1} such that $X_{i+1}^T X_{i+1} = I_r$ and R_{i+1} is upper triangular and

$$Z_{i+1} = X_{i+1} R_{i+1}$$

(If Z_{i+1} is rank deficient, choose the columns of X_{i+1} so that $X_{i+1}^T X_j = 0$ for $j \le i$.)

2.c Compute AX_{i+1} and $M_{i+1} = X_{i+1}^T AX_{i+1}$.

Step 2.b can be accomplished by using the Gram-Schmidt procedure or Householder method. In the absence of roundoff error the matrices X_i would satisfy

$$(2.1)\qquad \begin{aligned} X_i^T X_j &= 0 \quad \text{when} \quad i \neq j \\ X_i^T X_i &= I_r \end{aligned}$$

In order to guarantee that (2.1) be satisfied, all the matrices X_i must be forced by an orthogonalization proceed to be orthogonal to the previously generated matrices. Complete details of the algorithm are given in [6] and [7].

Thus a matrix J is generated such that

$$J = \begin{bmatrix} M_1 & R_2^T & & & \bigcirc \\ R_2 & M_2 & \cdot & \cdot & \\ & \cdot & \cdot & \cdot & \\ & & \cdot & \cdot & R_\ell^T \\ \bigcirc & & & R_\ell & M_\ell \end{bmatrix} .$$

Since the $\{R_i\}_{i=2}^{\ell}$ are upper triangular matrices, the matrix J is a band matrix with

$$j_{gh} = 0 \quad \text{when} \quad |g - h| \geq r + 1 \ .$$

Now if

$$J = Q \Lambda Q^T$$

where Λ is the diagonal matrix of eigenvalues of J and Q is the orthogonal matrix of eigenvectors then

$$\Lambda = Q^T J Q \ .$$

Hence, the matrix Q corresponds to the matrix X^T. Indeed if X_1 equals the first r rows of the matrix Q then

$$X^T = QD$$

where D is a diagonal matrix with $d_i = \pm 1$.

Thus, in order to generate a five diagonal matrix K, it is necessary to compute the first two components of the eigenvectors of the matrix. Using the Lemma given in section 1, we are able to compute the first column of the $n \times 2$ matrix of X_1 and in the next section we show how the second column of X_1 is computed. Once these two columns are determined, we are able to generate a five diagonal matrix by using the block Lanczos algorithm. The general problem is solved by determining the $n \times (p-1)$ matrix X_1.

3. Generation of the Five Diagonal Matrix

As was pointed out in the previous section, it is possible to construct the five diagonal matrix K, using the block Lanczos algorithm if the first two components of the eigenvectors of K are known. Using the Lemma of section 1, we are able to construct the square of the first component of the eigenvectors of K from the eigenvalues $\{\kappa_i\}_{i=1}^{n}$, $\{\lambda_i\}_{i=1}^{n-1}$ and the square of the first component of the eigenvectors of L from the eigenvalues $\{\lambda_i\}_{i=1}^{n-1}$, $\{\mu_i\}_{i=1}^{n-2}$. We now show how the first component of the eigenvectors of L determines the second component of the eigenvectors of K.

Let us write

$$K = \left(\begin{array}{c|c} a_1 & b_1, c_1, 0 \ldots\ 0 \\ \hline b_1 & \\ c_1 & \\ 0 & L \\ \vdots & \\ 0 & \end{array}\right) \equiv \left(\begin{array}{c|c} a_1 & \underset{\sim}{b}^T \\ \hline \underset{\sim}{b} & L \end{array}\right) . \tag{3.1}$$

The matrix of eigenvectors of K are denoted as Q and the matrix of eigenvectors of L as P so that

$$K\underset{\sim}{q}_j = \kappa_j \underset{\sim}{q}_j \quad (j = 1,2,\ldots,n) \quad \text{and} \quad L\underset{\sim}{p}_j = \lambda_j \underset{\sim}{p}_j \quad (j = 1,2,\ldots,n-1) .$$

It is useful to partition the eigenvector $\underset{\sim}{q}_j$ so that

$$\underset{\sim}{q}_j = \begin{pmatrix} x_j \\ \underset{\sim}{y}_j \end{pmatrix} \tag{3.2}$$

where x_j is a scalar and $\underset{\sim}{y}_j$ is a vector with $(n-1)$ components. We have that

$$x_j = q_{1j} \ ;$$

we have already computed $x_j^2 = q_{1j}^2$ and p_{1j}^2. We need y_{1j} in order to construct the five diagonal matrix K.

We examine the relationship between the eigenvalues of K and L and the vector $\underset{\sim}{b}$. Using a simple argument, it can be shown that given $\{\kappa_i\}_{i=1}^{n}$ and $\{\lambda_i\}_{i=1}^{n-1}$ and if

$$\hat{\underset{\sim}{b}} = P^T \underset{\sim}{b}$$

then

(3.3)
$$\hat{b}_k^2 = - \frac{\prod_{i=1}^{n} (\kappa_i - \lambda_k)}{\prod_{\substack{i=1 \\ i \neq k}}^{n} (\lambda_i - \lambda_k)} .$$

Since $K\underset{\sim}{q}_j = \kappa_j \underset{\sim}{q}_j$,

$$\underset{\sim}{b} x_j + L\underset{\sim}{y}_j = \kappa_j \underset{\sim}{y}_j$$

so that

$$(L - \kappa_j I)\underset{\sim}{y}_j = -\underset{\sim}{b} x_j .$$

Using the decomposition $L = P \Lambda P^T$, we have

$$\underset{\sim}{y}_j = -P(\Lambda - \kappa_j I)^{-1} P^T \underset{\sim}{b} x_j$$

$$= -P(\Lambda - \kappa_j I)^{-1} \hat{\underset{\sim}{b}} q_{1j} .$$

Thus

(3.4)
$$q_{2j} \equiv y_{1j} = - \sum_{k=1}^{n-1} \frac{p_{1k}\hat{b}_k}{(\lambda_k - \kappa_j)} q_{1j} .$$

Therefore to compute the five diagonal matrix K, one proceeds as follows.

Algorithm

1. Compute q_{1j}^2 and p_{1j}^2 by (1.5).
2. Compute $\hat{b}_j^2$ by (3.3).
3. Compute q_{2j} by (3.4).
4. Apply the block Lanczos Algorithm with

$$X_1 = \begin{pmatrix} q_{11} & q_{21} \\ q_{12} & q_{22} \\ \vdots & \vdots \\ q_{1n} & q_{2n} \end{pmatrix} .$$

Note that the sign of $p_{1k} \times \hat{b}_k$ and q_{1j} is not determined. Indeed for different sign configurations, one would expect to get different matrices K, and this has been verified experimentally. The (1,1) and (2,2) elements of the matrix (a_1 and a_2) are uniquely determined since

$$\text{tr } K = \sum_{i=1}^{n} \kappa_i , \quad \text{tr } L = \sum_{i=1}^{n-1} \lambda_i , \quad \text{tr } M = \sum_{i=1}^{n-2} \mu_i$$

and hence

$$a_1 = \sum_{i=1}^{n} \kappa_i - \sum_{i=1}^{n-1} \lambda_i , \qquad a_2 = \sum_{i=1}^{n-1} \lambda_i - \sum_{i=1}^{n-2} \mu_i .$$

Finally, the signs of the elements c_i $(i = 1,2,\ldots,n-1)$ and b_i $(i = 2,4,\ldots,n-2)$ will not be determined uniquely since they depend upon an arbitrary choice of signs in the orthogonalization procedure used in generating the matrix X_{i+1} from Z_{i+1} in the block Lanczos algorithm.

We have primarily discussed the generation of the five diagonal case but the ideas given here are valid for determining a matrix of any specified band width. We have developed programs which produce excellent numerical results using the above described techniques. An important feature of the algorithm is that the elements of the matrix X_1 are developed by a recursive procedure so that a high level language such as ALGOL W is particularly useful for the solution of such problems. We shall describe the more general situation and give programs in a future paper.

Acknowledgement

The authors wish to thank Professor Victor Barcilon of the University of Chicago for suggesting this problem and for his helpful comments.

References

[1] Harry Hochstadt, "On the construction of a Jacobi matrix from spectral data," Linear Algebra Appl. 8 (1974) 435-446.

[2] Ole H. Hald, "Inverse eigenvalue problems for Jacobi matrices," Linear Algebra Appl. 14 (1976) 63-86.

[3] C. de Boor and G. H. Golub, "The numerically stable reconstruction of a Jacobi matrix from spectral data," to be published in Linear Algebra Appl. (available as MRC Technical Summary Report #1727).

[4] G. H. Golub, "Some modified matrix eigenvalue problems," SIAM Review 15 No. 2 (1973) 318-334.

[5] G. H. Golub, "Some uses of the Lanczos algorithm in numerical linear algebra" in Topics in Numerical Analysis, John J. H. Miller (ed.) Academic Press, Inc. (1973) 173-184.

[6] R. R. Underwood, "An iterative block Lanczos method for the solution of large sparse symmetric eigenproblems, Ph.D. dissertation, Stanford University, Stanford, California.

[7] G. H. Golub and R. R. Underwood, "The block Lanczos method for computing eigenvalues," Proceedings of the Symposium on Mathematical Software Madison, 1977. Academic Press.

MULTIVARIATE APPROXIMANTS WITH BRANCH POINTS

J. S. R. Chisholm

1. Introduction

In 1973, I published a paper[1] introducing a 2-variable generalisation of diagonal Padé approximants. In a series of papers, various generalisations of this type of rational approximant were proposed: N-variable diagonal approximants[2], 2-variable simple off-diagonal approximants[3], simple and general off-diagonal N-variable approximants[4], known as S.O.D.'s and G.O.D.'s, and rotationally covariant diagonal and simple off-diagonal approximants[5]. The original definition[1,6] was framed so that the approximants satisfied a number of "basic properties", and the structure of the algebraic properties of the equations[7] proved to be both computationally and algebraically good. The various generalisations[2-5] were defined to preserve, as far as could be expected, both the set of basic properties and the desirable algebraic structure.

These N-variable approximants are all rational functions of several complex variables, and hence are single-valued. As with Padé approximants, it is possible to represent branch point singularities by singularities of the rational approximants, but this is not a very efficient procedure[8,9]. In principle it would be better to use approximants which themselves possess branch points. For functions of one variable, Shafer[10] introduced another generalisation of Padé approximants which satisfies a quadratic or higher power algebraic equation, defining approximants with quadratic or higher order branch points. In calculating fourth order quantum field theory matrix elements, functions possessing an infinity of Riemann sheets, Shafer's quadratic and cubic give at least two figures greater accuracy than the corresponding Padé approximants[11].

The N-variable approximants I am going to describe are a generalisation both of the N-variable approximants and of Shafer's approximants. The work only began after Easter, 1977, so that it is too early to discuss the effectiveness of the approximation method; Leslie Short has been writing programmes to calculate the approximants, and has some preliminary results. Since the definition of the new approximants depends on the properties of both the N-variable rational approximants and on Shafer's approximants, I begin by summarising these.

2. Shafer's Approximants

Given a formal power series in one variable

$$f(z) = \sum_{\gamma=0}^{\infty} c_\gamma z^\gamma \quad , \tag{2.1}$$

the [m/n] Padé approximant is defined by constructing two polynomials $M(z)$, $N(z)$, of degrees m,n respectively, satisfying the formal identity

$$N(z)\ f(z) - M(z) = O(z^{m+n+1}) \quad . \tag{2.2}$$

Then the approximant $f_{m/n}(z)$ is given by

$$N(z)\ f_{m/n}(z) - M(z) = 0 \quad . \tag{2.3}$$

Normally, $f_{m/n}(z)$ is unique.

The Shafer quadratic approximants of (2.1) are defined in terms of three polynomials $P(z)$, $Q(z)$, $R(z)$, of degrees ℓ,m,n respectively, satisfying the formal identity

$$P(z)\ f^2(z) + Q(z)\ f(z) + R(z) = O(z^{\ell+m+m+2)} \quad , \tag{2.4}$$

where $f^2(z)$ is the formal square of the series (2.1). Normally, this determines uniquely the ratios of the (ℓ+m+n+3) coefficients in P,Q,R . The approximant $f_{\ell/m/n}(z)$ is then the solution of

$$P(z)\ [f_{\ell/m/n}(z)]^2 + Q(z)\ f_{\ell/m/n} + R(z) = 0 \quad . \tag{2.5}$$

This defines a function on two Riemann sheets; the sheet which corresponds to the series (2.1) has

$$f_{\ell/m/n}(0) = c_0 \quad . \tag{2.6}$$

One coefficient, say $p_0 \equiv P(0)$, can be arbitrarily chosen. In the definition of the new approximants, it is sometimes useful to consider both p_0 and $q_0 \equiv Q(0)$ to be given, replacing (2.4) by the formal identity

$$P(z)\ f^2(z) + Q(z)\ f(z) + R(z) = O(z^{\ell+m+n+1}) \quad . \tag{2.7}$$

This makes little difference for approximants in one variable, since we can always determine $p_0 : q_0$ by equating to zero the coefficient of $z^{\ell+m+n+1}$ if we wish. Alternative methods of choosing this ratio are possible: for example, if the value $f_1(0)$ of $f(z)$ at the origin on a second Riemann sheet were known, we could add to (217) the equation

$$p_0\ [f_1(0)]^2 + q_0\ f_1(0) + r_0 = 0 \quad , \tag{2.8}$$

where $r_0 = R(0)$.

3. N-variable Rational Approximants

I shall discuss the original 2-variable diagonal approximants, since their properties exemplify the whole set of rational approximants. Given a double power series

$$f(\underset{\sim}{z}) = \sum_{\underset{\sim}{\gamma}=\underset{\sim}{0}}^{\underset{\sim}{\infty}} c_{\underset{\sim}{\gamma}} \, \underset{\sim}{z}^{\underset{\sim}{\gamma}} , \qquad (3.1)$$

where

$$\underset{\sim}{z}^{\underset{\sim}{\gamma}} = z_1^{\gamma_1} z_2^{\gamma_2} ,$$

the diagonal approximants are of the form

$$f_{\underset{\sim}{m}/\underset{\sim}{m}}(\underset{\sim}{z}) = \frac{\sum_{\underset{\sim}{\alpha}=\underset{\sim}{0}}^{\underset{\sim}{m}} a_{\underset{\sim}{\alpha}} \underset{\sim}{z}^{\underset{\sim}{\alpha}}}{\sum_{\underset{\sim}{\beta}=\underset{\sim}{0}}^{\underset{\sim}{m}} b_{\underset{\sim}{\beta}} \underset{\sim}{z}^{\underset{\sim}{\beta}}} , \qquad (3.2)$$

the summation being over the square S_1 of Fig.1. The ratios of $\{a_{\underset{\sim}{\alpha}}\}$, $\{b_{\underset{\sim}{\beta}}\}$ are determined by equating to zero certain combinations of coefficients in

$$E(\underset{\sim}{z}) \equiv \sum_{\underset{\sim}{\varepsilon}=\underset{\sim}{0}}^{\underset{\sim}{\infty}} e_{\underset{\sim}{\varepsilon}} \underset{\sim}{z}^{\underset{\sim}{\varepsilon}}$$

$$= \left[\sum_{\underset{\sim}{0}}^{\underset{\sim}{m}} a_{\underset{\sim}{\alpha}} \underset{\sim}{z}^{\underset{\sim}{\alpha}}\right] \left[\sum_{\underset{\sim}{0}}^{\underset{\sim}{\infty}} c_{\underset{\sim}{\gamma}} \underset{\sim}{z}^{\underset{\sim}{\gamma}}\right] - \sum_{\underset{\sim}{0}}^{\underset{\sim}{m}} b_{\underset{\sim}{\beta}} \underset{\sim}{z}^{\underset{\sim}{\beta}} . \qquad (3.3)$$

The equations are

$$e_{\underset{\sim}{\varepsilon}} = 0 \qquad (\varepsilon_1+\varepsilon_2 \leqslant 2m) , \qquad (3.4)$$

over the region $S_1+S_2+S_3$ in Fig.1, and the "symmetrised equations"

$$e_{\varepsilon,2m+1-\varepsilon} + e_{2m+1-\varepsilon,\varepsilon} = 0 \qquad (1 \leqslant \varepsilon \leqslant m) \qquad (3.5)$$

corresponding to the line S_4 in Fig.1. Weight factors can be introduced in (3.5), but I shall not discuss these. The form (3.2) of the approximant and the system of equations (3.4)-(3.5) was chosen to satisfy the properties:

(i) Symmetry between the variables.

(ii) Projection: $f_{\underset{\sim}{m}/\underset{\sim}{m}}(z_1,0)$ is the [m/m] Padé approximant to the series $f(z_1,0)$.

(iii) Reciprocal covariance: The $[\underset{\sim}{m}/\underset{\sim}{m}]$ approximant to $f^{-1}(\underset{\sim}{z})$ is $[f_{\underset{\sim}{m}/\underset{\sim}{m}}(\underset{\sim}{z})]^{-1}$.

(iv) Homographic covariance: the [m/m] approximant to the expansion of

$f[Aw_1/(1+B_1w_1) , Aw_2/(1+B_2w_2)]$ is $f_{\underset{\sim}{m}/\underset{\sim}{m}} [Aw_1/(1+B_1w_1) , Aw_2/(1+B_2w_2)]$.

The last two properties reduce[1,2,4] to the

Rectangle Rule: If any point $\underset{\sim}{\varepsilon}$ contributes to the set of equations (3.4)-(3.5), then all points

$$\{\underset{\sim}{\alpha} \mid \alpha_i \leq \varepsilon_i \quad (i=1,2) \quad ; \quad \underset{\sim}{\alpha} \neq \underset{\sim}{\varepsilon}\}$$

contribute unsymmetrised equations of form (3.4).

The full structure of the equations (3.4)-(3.5) was expressed in terms of L-shaped "prongs" by Hughes Jones and Makinson[7]. Prong 0, shown in Fig.1, consists of the two axes; the projection property ensures that (3.4) holds at points out to 2m on each axis, and at no other points on prong 0. Prong σ, with vertex (σ,σ), introduces $4(m-\sigma)+2$ new variables; to determine these, there are $4(m-\sigma)+1$ equations (3.4) and one symmetrised equation (3.5). This equality of numbers of equations and new variables ensures that the matrix of the system (3.4)-(3.5) is of block lower-diagonal form, with square blocks on the diagonal. These square blocks are closely related to the matrices occurring in the calculation of Padé approximants from the one-variable series $f(z_1,0)$ and $f(0,z_2)$.

The N-variable and off-diagonal generalisations of the 2-variable diagonal approximants satisfy the basic properties and have an analogous algebraic structure based on prongs.

4. N-variable t-power Approximants

I shall discuss the definition and properties of 2-variable t power off diagonal approximants in detail, and then comment briefly on the N-power generalisation, and on diagonal approximants.

Given a double power series

$$f(\underset{\sim}{z}) = \sum_{\underset{\sim}{\gamma}=\underset{\sim}{0}}^{\infty} c_{\underset{\sim}{\gamma}} \, \underset{\sim}{z}^{\underset{\sim}{\gamma}} , \tag{4.1}$$

we proceed to define (t+1) polynomials

$$P^{(k)}(\underset{\sim}{z}) \equiv \sum_{\underset{\sim}{\alpha}\subset S_k} p_{\underset{\sim}{\alpha}}^{(k)} \, \underset{\sim}{z}^{\underset{\sim}{\alpha}} \qquad (k=0,1,\ldots,t) \tag{4.2}$$

with coefficients lying in rectangular regions

$$S_k = \{\underset{\sim}{\alpha} \mid 0 \leq \alpha_i \leq m_{i;k} \quad ; \quad i=1,2\} \tag{4.3}$$

on the integer lattice. The formal k^{th} power of (4.1) is

$$f^k(\underset{\sim}{z}) \equiv \sum_{\underset{\sim}{\alpha}=\underset{\sim}{0}}^{\infty} c_{\underset{\sim}{\gamma}}^{(k)} \, \underset{\sim}{z}^{\underset{\sim}{\gamma}} . \tag{4.4}$$

We define the (formal) series

$$E(\underset{\sim}{z}) = \sum_{k=1}^{t} P^{(k)}(\underset{\sim}{z})\, f^{k}(\underset{\sim}{z}) + P^{(0)}(\underset{\sim}{z}) \quad . \tag{4.5}$$

Given a lattice point $\underset{\sim}{\varepsilon}$, define the rectangular lattice region

$$S_{\underset{\sim}{\varepsilon}} = \{\underset{\sim}{\alpha} \mid 0 \leq \alpha_i \leq \varepsilon_i \quad ; \quad i=1,2\} \quad . \tag{4.6}$$

Then the coefficient of $\underset{\sim}{z}^{\underset{\sim}{\varepsilon}}$ in (4.5) is

$$e_{\underset{\sim}{\varepsilon}} = \sum_{k=1}^{t} \left[\sum_{\underset{\sim}{\alpha} \subset S_k \cap S_{\underset{\sim}{\varepsilon}}} p_{\underset{\sim}{\alpha}}^{(k)} \; c_{\underset{\sim}{\varepsilon}-\underset{\sim}{\alpha}}^{(k)} \right] + p_{\underset{\sim}{\varepsilon}}^{(0)} \; \delta(\underset{\sim}{\varepsilon} \subset S_0) \quad , \tag{4.7}$$

where

$$\delta(\underset{\sim}{\varepsilon} \subset S_0) = \left\{ \begin{array}{lcl} 1 & , & \underset{\sim}{\varepsilon} \subset S_0 \\ 0 & , & \varepsilon \not\subset S_0 \end{array} \right\} \quad . \tag{4.8}$$

The ratios of the coefficients $\{p_{\underset{\sim}{\alpha}}^{(k)}\}$ will be defined by the correct number of equations linear in $\{e_{\underset{\sim}{\varepsilon}}\}$. When $\{p_{\underset{\sim}{\alpha}}^{(k)}\}$ are determined, the approximant $f(\underset{\sim}{z};S_\ell)$ is determined by

$$\sum_{k=1}^{t} P^{(k)}(\underset{\sim}{z})\, [f(\underset{\sim}{z}\, ;\, S_\ell)]^k + P^{(0)}(\underset{\sim}{z}) = 0 \quad . \tag{4.9}$$

This defines a function with t Riemann sheets. The sheet corresponding to the series (4.1) is identified by ensuring that $f(\underset{\sim}{0};S_\ell) = c_0$, and if necessary by identifying some first derivatives.

The linear equations are determined by use of L-shaped prongs; nearly all equations are of the form (3.4), but some prongs need a symmetrised equation (3.5) from the two end-points. The prong method ensures that the equations are of block lower-diagonal form, with square blocks on the diagonal. The fact that the prongs are L-shaped ensures that the basic properties are <u>almost</u> satisfied. In Fig.2 we give an example of the prong structure for a cubic 2-variable approximant; there are four regions $\{S_k\}$, taken to be of dimensions (5,2) , (3,6) , (6,5) , (7,8) . The prongs have vertices of the form (σ,σ) . In general, they are grouped into sets of adjacent prongs, depending on the number of $\{S_k\}$ which contain the prong vertex. In Fig.2, prongs σ=7,6 have vertices in only one of $\{S_k\}$; these prongs fill out a rectangle in the corner of this S_k , and the edges of the region have "slope 0" . No symmetrised equations are needed for these prongs. Prongs σ=5,4 have vertices in two of $\{S_k\}$; the ends of the prongs lie on lines of "slope 1" , and the two points at the end of each prong contribute a symmetrised equation. The next prong σ=3 has vertex in three of $\{S_k\}$, and no symmetrised equation is needed; if there were more than one such prong, the end-points would lie on lines of "slope 2" . The prongs σ=2,1 lie in all four S_k ; their end-points lie on lines of "slope 3", and correspond to symmetrised equations. In general, end-points of prongs lying in r of the rectangles $\{S_k\}$ lie on lines of "slope" (r-1) ; the end-points correspond to symmetrised equations only when r is even, not when r is odd.

Prong 0 needs some discussion. If we assume that

$$\{\underset{\sim}{p}_0^{(k)} \quad ; \quad k=1,2,\ldots,t\} \tag{4.10}$$

are given, and only $\underset{\sim}{p}_0^{(0)}$ of the zero-order coefficients is to be determined, then the correct number of equations are obtained by writing $e_{\underset{\sim}{\varepsilon}} = 0$ for points $\underset{\sim}{\varepsilon}$ on the two axes out to the points with coordinates

$$\sum_{k=0}^{t} m_{i;k} \qquad (i=1,2) \quad . \tag{4.11}$$

One way of determining the (t-1) ratios of (4.10) is to obtain (t-1) further equations from points $\underset{\sim}{e}$ on the axes. If the power t is odd, we add $\frac{1}{2}$(t-1) points on each axis; the approximants then project to Shafer's approximants when we put z_2=0 or z_1=0 . But if t is even, we have to add $\frac{1}{2}$(t-2) points on each axis, and include the symmetrised equations from the next pair of points; the symmetrised equation spoils the projection property to Shafer approximants.

An alternative way of determining the (t-1) ratios of (4.10) is to use the (different) values $f_1(0),\ldots,f_{t-1}(0)$ of the given function at the origin on (t-1) other Riemann sheets, if these are known, giving

$$\sum_{k=1}^{t} \underset{\sim}{p}_0^{(k)} \, [\underset{\sim}{f}_r(0)]^k + \underset{\sim}{p}_0^{(0)} = 0 \qquad (r=1,\ldots,t-1) \quad , \tag{4.12}$$

a generalisation of (2.8). The system of equations then satisfies the projection property for all t .

For N=3 variables, the picture is very similar. The prong structure is more complicated; prongs based on points (σ,σ,σ) are again grouped into sets, depending on how many rectangular solids S_k they are contained in; the groups again have prong lengths with "interval" 0,1,2,...,t-1; the end points of these groups of prongs contribute, in turn, no symmetrised equations, symmetrised equations of form

$$e_{\underset{\sim}{\varepsilon}_1} + e_{\underset{\sim}{\varepsilon}_2} + e_{\underset{\sim}{\varepsilon}_3} = 0 \quad , \tag{4.13}$$

and pairs of symmetrised equations of form

$$e_{\underset{\sim}{\varepsilon}_1} = e_{\underset{\sim}{\varepsilon}_2} = e_{\underset{\sim}{\varepsilon}_3} \quad . \tag{4.14}$$

For N(>3) variables, it is not in general possible to define a symmetrical set of equations; we have, for example, to construct two linear equations from four expressions $\{e_{\underset{\sim}{\varepsilon}}\}$.

For N(>2) variables, it is not always possible to ensure the full projection property to Shafer approximants unless equations (4.12) can be used. In general, the approximants satisfy reciprocal covariance.

For diagonal t-power N-variable approximants, the regions S_k $(k=0,1,\ldots,t)$ are identical hypercubes. Leslie Short has pointed out that the linear equations given here then have to be modified on the final prong, since they are otherwise inconsistent. There are several possible modifications, some of which affect the basic properties to some extent.

The t-power approximants have been defined and studied in two recent papers[12,13].

REFERENCES

1. Chisholm, J. S. R., Math.Comp.27,841 (1973).
2. Chisholm, J. S. R. and McEwan, J., Proc.Roy.Soc.A336,421 (1974).
3. Graves-Morris, P. R., Hughes Jones, R. and Makinson, G. J., J.I.M.A. 13,311 (1974).
4. Hughes Jones, R., J.Approx.Th.16,3 (1976).
5. Chisholm, J. S. R. and Roberts, D. E., Proc.Roy.Soc.A351,585 (1976).
6. Chisholm, J. S. R., Proceedings of 1976 Tampa Conference on Rational Approximation, ed. E. B. Saff, Academic Press (1977).
7. Hughes Jones, R. and Makinson, G. J., J.I.M.A. 13,299 (1974).
8. Roberts, D. E., Wood, D. W. and Griffiths, H. P., J.Phys.A8,9 (1975).
9. Roberts, D. E., submitted for publication.
10. Shafer, R. E., S.I.A.M.J.Num.Analysis 11,417 (1974).
11. Chisholm, J. S. R. and Short, L., Proceedings of 1977 St. Maximin Conference on Advanced Computational Methods in Theoretical Physics, ed. A. Visconti (Marseilles).
12. Chisholm, J. S. R., "Multivariate Approximants with Branch Points I", Proc.Roy. Soc., to be published.
13. Chisholm, J. S. R., "Multivariate Approximants with Branch Points II", submitted to Proc.Roy.Soc.

Fig.1.

Regions contributing equations for diagonal 2-variable rational approximants

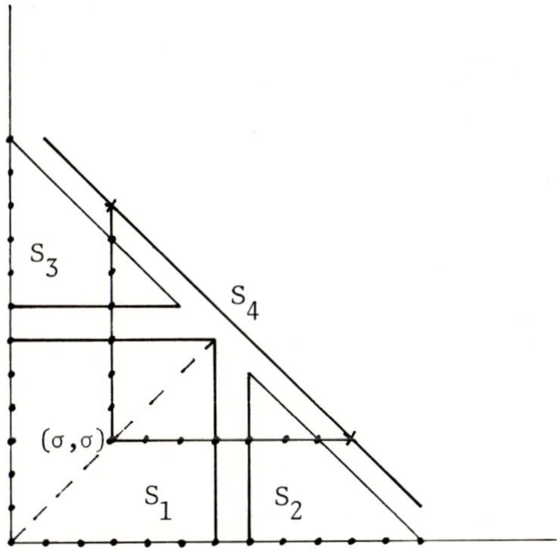

Prong lines —•—•—•—•—•—•—•—

Symmetrised points ×

Fig.2.

Prong structure for cubic (t=3) 2-variable approximants with rectangles $\{S_k\}$ of dimensions (5,2) , (3,6) , (6,5) , (7,8) .

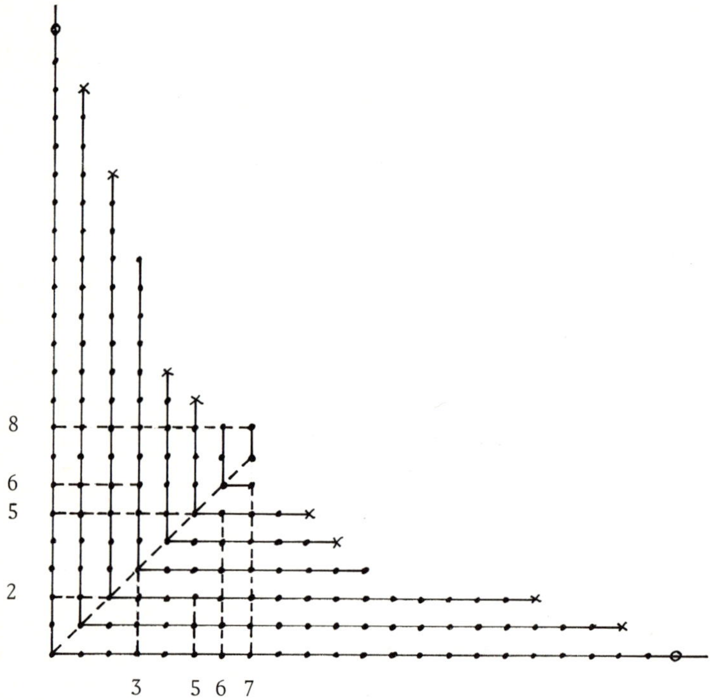

Prong lines

Symmetrised points ×

Optional points (prong 0) o

THE NUMERICAL TREATMENT OF SOME SINGULAR BOUNDARY VALUE PROBLEMS

Lothar Collatz

Summary: Every numerical method for getting approximate solutions of a singular boundary value problem should take care of the type of the occurring singularities, otherwise the convergence would be very slow. A list of singularities may be helpful for the treatment of concrete problems. Using approximation methods one can get in certain cases exact error bounds for the approximate solutions; this is illustrated by different examples, especially a free boundary value problem, for which an exact error bound for the free boundary is given.

I. Some general remarks

The singularities occurring in boundary value problems may be classified very roughly in the following way:

1. Singularities of geometric type
   a. In the finite domain, f.i. corners.
   b. Unbounded domains
2. Singularities of analytic type
   a. harmless } singularities of the coefficients
   b. serious } occurring in the Differential Equations
   c. coming from the differential equations (often "moving singularities")
   d. coming from the problem itself, f.i. free boundaries.

The methods used for numerical calculation are often:

<table>
<tr><td rowspan="2">Discretization methods</td><td colspan="2">Difference methods</td></tr>
<tr><td rowspan="2">Variational methods</td><td>Finite Element method, Splines</td></tr>
<tr><td rowspan="2">Parametric methods</td><td>Ritz-type methods</td></tr>
<tr><td colspan="2">Approximation methods</td></tr>
</table>

Using any of these methods it is very important to look carefully on the type of singularity otherwise one has to expect a very slow convergence of the numerical procedure.

There is no strong distinction between the different methods.

There are special methods using finite elements developped for the case of singularities, see f.i. Whiteman [73], Mitchell-Wait [77] a.o. In this lecture the approximation methods will be considered preferable because these methods are in certain cases the only ones which give exact error bounds for approximate solutions of the boundary value problem.

We consider only some simple examples for brevity and clearness, some of them for linear problems, but of course nonlinear problems have also been treated extensively.

II. Singularities of geometric type

Let us consider the torsion-problem for a beam the cross-section of which is a rhomb B in the x-y-plane with the size s=1, with the angle $\alpha=\frac{1}{3}\pi$ and the origin x=y=0 as centre, fig. 1. The function u(x,y) has to satisfy the Laplace-Equation

(2.1) $\Delta u=\frac{\partial^2 u}{\partial x^2}+\frac{\partial^2 u}{\partial y^2}=0$ in B

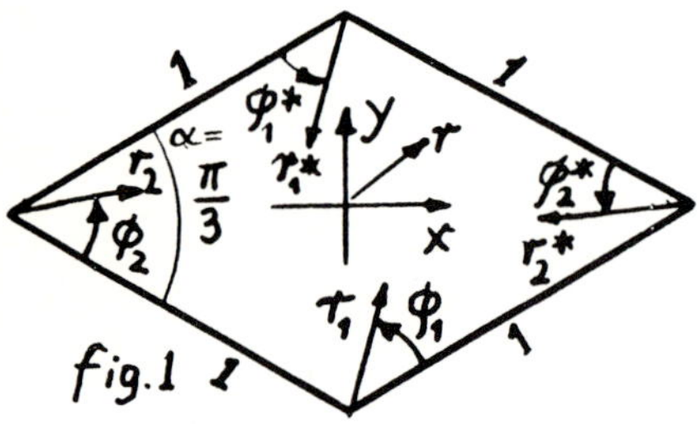
fig.1

and the boundary condition

(2.2) $u=r^2=x^2+y^2$ on ∂B.

We look at an approximate solution w for u in the form

(2.3) $u\approx w=\sum_{\nu=1}^{p} a_\nu w_\nu(x,y)$

where the w_ν satisfy the differential equation $\Delta w_\nu=0$ ($\nu=1,\dots,p$).

The error $\varepsilon=w-u$ can be bounded by the classical maximumprinciple: Calculating an error bound on ∂B the same bound holds in the whole domain B:

(2.4) From $|\varepsilon|\leq K$ on ∂B follows $|\varepsilon|\leq K$ in B.

One can choose for the w_ν harmonic polynomials like

(2.5) $1, x, y, xy, x^2-y^2, x^3-3xy^2, \dots$

or one can take care of the singularities and use (with the angles ϕ_j, ϕ_j^* and the distances r_j, r_j^* (j=1,2) as in figure 1)

$$(2.6)\begin{cases} v_1=r_1^{3/2}\sin\frac{3\phi_1}{2}+r_1^{*3/2}\sin\frac{3\phi_1^*}{2} \\ v_2=r_2^{3}\sin(3\phi_2)+r_2^{*3}\sin(3\phi_2^*) \end{cases}$$

In each of these cases one has to determine the constants a_ν in (2.3) in such a way that on the boundary the prescribed values of u are approximated as good as possible. This is a classical problem of linear Tschebyscheff Approximation for which routines f.i. the Remez-algorithm are available. One gets the following exact error bounds:

Using	$\lvert\varepsilon\rvert = \lvert w-u\rvert \leq$
$w_1=1$	0,28
polynomials up to degree 2 (included)	0.1
$w_2=v_1$, $w_3=v_2$	0.0055

The error bound by using the singular terms in (2.6) gives better results than that only using the quadratic polynomials from (2.5). Other examples, f.i. the ideal flow of a liquid over a threshold in an unbounded domain, fig. 2, are described in Collatz [73] p.11:

$$\Delta u=0 \text{ in } B \ (-\infty<x<\infty, \psi(x)=\frac{1}{1+x^2}<y<\infty)$$

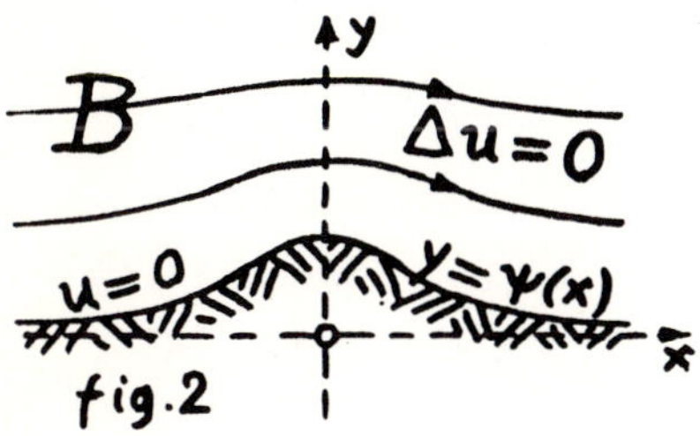

fig.2

$u=0$ for $y=\psi(x)$, $\lim_{y\to\infty} [u(x,y)-y]=0$ for fixed x.

We take as approximate solution

$$u\approx w=y-\sum_{\nu=1}^{p} a_\nu \frac{y+c_1}{x^2+(y+c_2)^2} .$$

For $p=2$, $a_1=0.79801$, $a_2=1.02523$, $c_1=6.3485$ one gets the exact error bound $\lvert w-u\rvert \leq 0.0624$. One can easily improve this bound by taking larger values for p.

It can be necessary in complicated cases to approximate in the interior of B and on the boundary (simultaneous approximation compare Bredendiek [70] [76]).

III. List of functions with singularities

It may be useful to have a list of singularities of simple functions. The following part of such a table gives for some functions u(x,y) the values of the functions and its first derivatives along the coordinate-axes and along a straight line through the origin; r,φ mean polarcoordinates

(3.1) $\quad x=r\cos\varphi, \quad y=r\sin\varphi$

TABLE Extract (for illustration): Behaviour of some special functions

u	$\frac{\partial u}{\partial x}$	$\frac{\partial u}{\partial y}$	on $x>0, y=0$: u	on $x>0, y=0$: $\frac{\partial u}{\partial y}$	on $x=0, y>0$: u	on $x=0, y>0$: $\frac{\partial u}{\partial x}$	on $x=-s<0, y=0$: u	on $x=-s<0, y=0$: $\frac{\partial u}{\partial y}$	on $x=r\cos\beta, y=r\sin\beta$: u	on $x=r\cos\beta, y=r\sin\beta$: $\frac{\partial u}{\partial \varphi}$
$\varphi=\mathrm{Im}(\ln z)$	$-y/r^2$	x/r^2	0	$1/x$	$\pi/2$	$-1/y$	π	$-1/s$	β	1
r	x/r	y/r	x	0	y	0	s	0	r	0
$\ln r=\mathrm{Re}(\ln z)$	x/r^2	y/r^2	$\ln x$	0	$\ln y$	0	$\ln s$	0	$\ln r$	0
$r^\alpha\cos(\alpha\varphi)=\mathrm{Re}(z^\alpha)$	RC	$-RS$	x^α	0					$r^\alpha\cos(\alpha\beta)$	$-\alpha r^\alpha\sin(\alpha\beta)$
$r^\alpha\sin(\alpha\varphi)=\mathrm{Im}(z^\alpha)$	RS	RC	0	$\alpha x^{\alpha-1}$					$r^\alpha\sin(\alpha\beta)$	$\alpha r^\alpha\cos(\alpha\beta)$
$\mathrm{Im}(z^\alpha\ln z)= r^\alpha\ln r\cdot\sin(\alpha\varphi)+ r^\alpha\varphi\cdot\cos(\alpha\varphi)$	$R[LS+\varphi C]$	$R[LC-\varphi S]$	0	$\frac{\alpha(\ln x+1)}{x}$						
....	.....	.....	.....	.....	.....					

$z=x+iy=re^{i\varphi}$; $R=\alpha r^{\alpha-2}$; $L=\ln r+\frac{1}{\alpha}$

$S=x\cdot\sin(\alpha\varphi)-y\cos(\alpha\varphi)$

$C=x\cdot\cos(\alpha\varphi)+y\sin(\alpha\varphi)$

The table was used in some of the examples here described. One can use a similar table for functions $u(x,y,z)$ in the three-dimensional space.

IV. Using higher singularity-terms.

It is necessary for more complicated problems to use monotonicity theorems instead of the principle of boundary maximum, especially for all nonlinear problems and even for linear problems in the case of more complicated boundary conditions. This may be illustrated for one example (Many different problems have been treated on the computer).

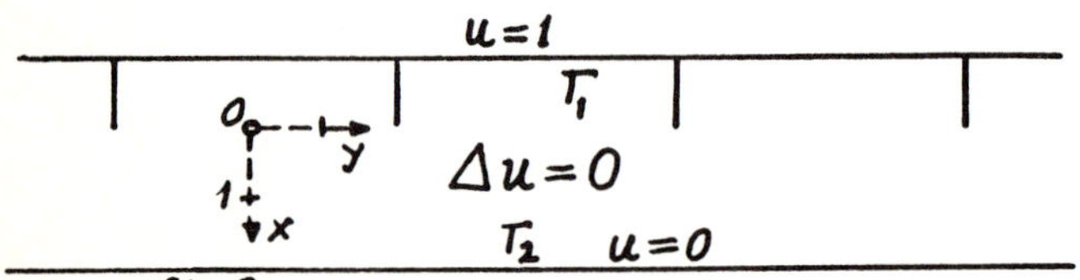

fig.3

We ask for a function $u(x,y)$ in the x-y-plane, which satisfies the Laplace equation (2.1) in the following domain B and the boundary conditions

$$(4.1)\quad \begin{cases} u=1 \text{ on } \Gamma_1 \\ u=0 \text{ on } \Gamma_2 \text{ or } x=2. \end{cases}$$

Γ_1 consists of the line $x=-1$ and the line segments

(4.2) $y=4k-2$ $(k=0, \pm1, \pm2, \ldots)$, $-1\leq x\leq 0$.

B is the strip $-1<x<2$ except the pieces (4.2), fig.3. Many physical interpretations can be given for $u(x,y)$, f.i. distribution of temperature, of a potential or $u=$const as streamlines of an ideal flow of a river, where the line segments (4.2) are breakwaters, fig. 4.

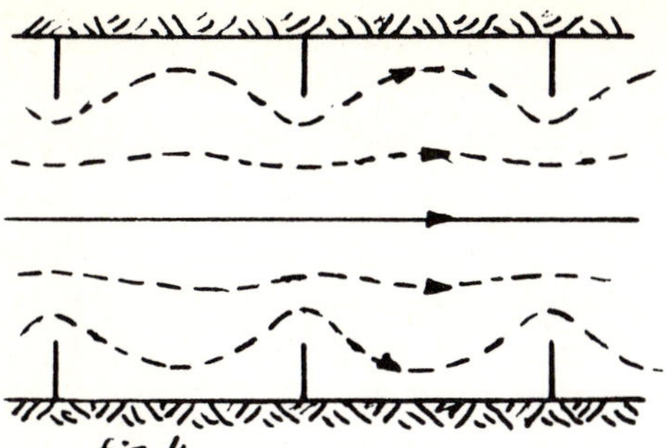

fig. 4

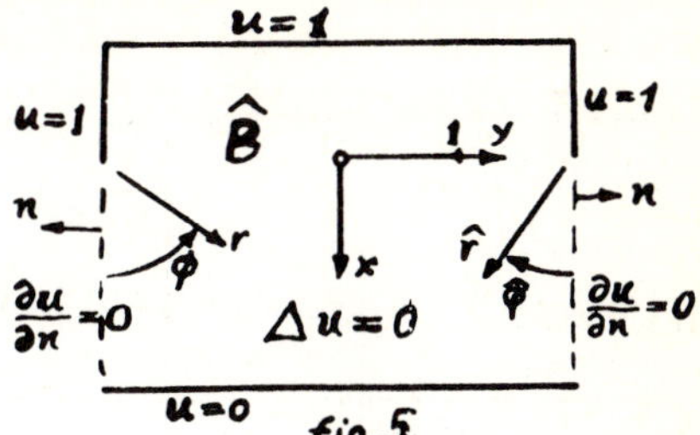

fig. 5

One can restrict oneself to the domain $\hat{B}$, which is the intersection of B with $|y|<2$ fig. 5. We introduce the Polarcoordinates $r,\phi,\hat{r},\hat{\phi}$ at the corners $x=0$, $y=\pm 2$, fig. 5, and the functions (according to the table in III)

$$v_1^* = r^{1/2}\cos(\tfrac{\phi}{2}) + \hat{r}^{1/2}\cos(\tfrac{\hat{\phi}}{2})$$

$$v_2^* = r^{3/2}\cos(\tfrac{3\phi}{2}) + \hat{r}^{3/2}\cos(\tfrac{3\hat{\phi}}{2})$$

Then (4.3) $w=a_1 v_1^* + a_2 v_2^* + \sum_{j=3}^{p} a_j v_j, \left[v_{k+3} = \mathrm{Re}(x+iy)^k (k=0,1,2,\ldots)\right]$

satisfies the Laplace-Equation $\Delta w=0$.

Here we use the following monotonicity principle:
The boundary of $\hat{B}$ contains two parts $\tilde{\Gamma}_1$, $\tilde{\Gamma}_2$; we have on $\tilde{\Gamma}_1$ prescribed values for the function u and on $\tilde{\Gamma}_2$ prescribed values for the normal derivative $\frac{\partial u}{\partial n}$ of u (with n as outer normal).

If $z(x,y)$ satisfies

$$(4.4)\quad \begin{cases} -\Delta z \geq 0 & \text{in } B \\ z \geq 0 & \text{on } \tilde{\Gamma}_1 \\ \frac{\partial z}{\partial n} \geq 0 & \text{on } \tilde{\Gamma}_2, \end{cases}$$

Then (4.4) has the consequence $z \geq 0$ in B.

Taking as function z the error $\varepsilon=w-u$, the monotonicity principle gives exact error bounds. We have to apply twice the onesided Tschebyscheff Approximation, once with certain values of the a_ν for a lower bound and another time with certain other values of the a_ν for an upper bound. The following table gives exact error bounds for 3 cases: working only with polynomials ($a_1=a_2=0$), working with the singularity v_1^* (that means $a_2=0$) and working also with the higher singularity v_2^*. It is seen from the table that without the terms with the singularity, that means working only with polynomials, the convergence is very bad, the error bound is by adding 1o polynomial-terms going down only from o.133 to o.o962; working with v_1^* the error bound decreases rapidly and even more rapidly by considering v_2^* too.

Exact error bounds $|w-u| \leqq$

Number of polynomials	without v_1^*, v_2^*	with v_1^*	with v_1^*, v_2^*
0	1	0.720	0.238
1	0.5	0.093 6	0.066 8
4	0.133	0.055 2	0.006 63
10	0.108	0.008 13	0.000 274
14	0.096 2	0.003 31	0.000 112

In cases like these one can use conformal mapping (here elliptic functions) to get exact solutions (Whiteman - Papamichael [72]) or to avoid the singularity (Whiteman [76]) or to get error bounds (Wetterling [76]) but this example was given to illustrate the influence of the singularities to the numerical computation.

V. Exact error bound for a free boundary value problem

Free boundary value problems occurred in the last decades in many different fields of applications (f.i. Baiocchi [74] a.o.)

Here we shall treat a simple and classical idealized case, the Stefan Problem of a melting ice-cover of a lake, situated in the half plane $x>0$ of an x-y-plane, whereas the halfplane $x<0$ is solid ground. At time t there is a strip $0<x<s(t)$ of free water at temperature $u=u(x,t)>0$, the rest $x>s(t)$ is covered with ice, fig. 6.

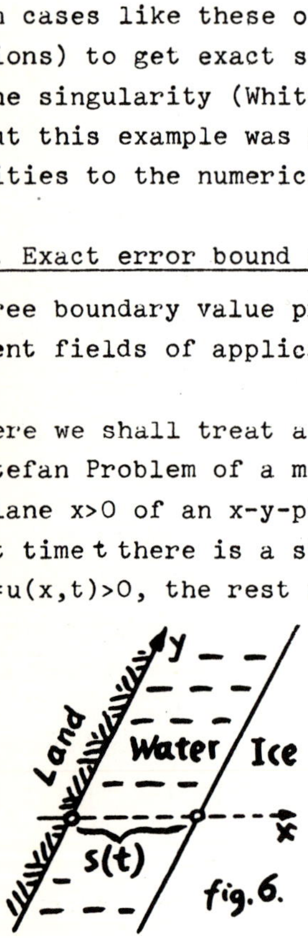

fig. 6.

Let B be the domain $0<t<T$, $0<x<s(t)$ in the x-t-plane fig. 7. The boundary B consists of parts $\Gamma_1, \Gamma_2, \Gamma_3, \Gamma_4$. We have for the unknown function $u(x,t)$ the heat-conduction equation

$$(5.1) \quad Lu = \frac{\partial u}{\partial t} - \frac{\partial^2 u}{\partial x^2} = 0 \quad \text{in } B$$

and the boundary conditions

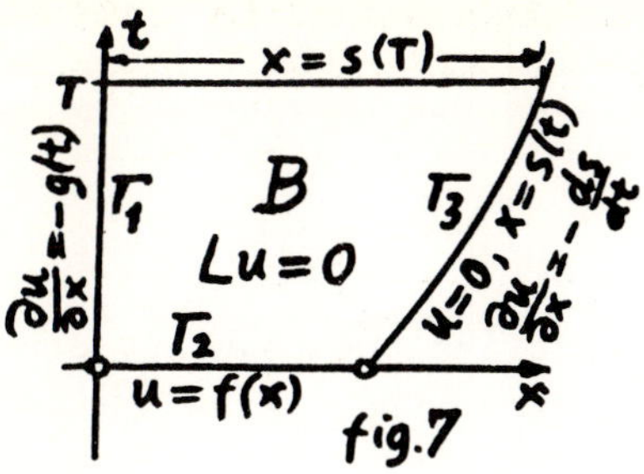

fig.7

(5.2) $\frac{\partial u}{\partial x} = - g(t)$ (heat entering from land) on Γ_1

(5.3) $u = f(x)$ (given distribution) on Γ_2

(5.4) $x = s(t)$, $u = 0$, $\frac{\partial u}{\partial x} = - \frac{ds}{dt}$ (on Γ_3 (heat balance at the free boundary.)

One wants to determine the free bondary $x=s(t)$.

The Gauss-Theorem gives for s(t) the nonlinear integral equation, (Evans [51])

$$(5.5)\quad s(T)=\Phi(s(T))=\int_0^T g(t)dt+\int_0^{s(0)} f(x)dx+s(0)-\int_0^{s(T)} \hat{u}(x,T,s(t))dx$$

where $\hat{u}(x,t,\sigma(t))$ is the solution of $L\hat{u}=0$ in the domain (fig.8)

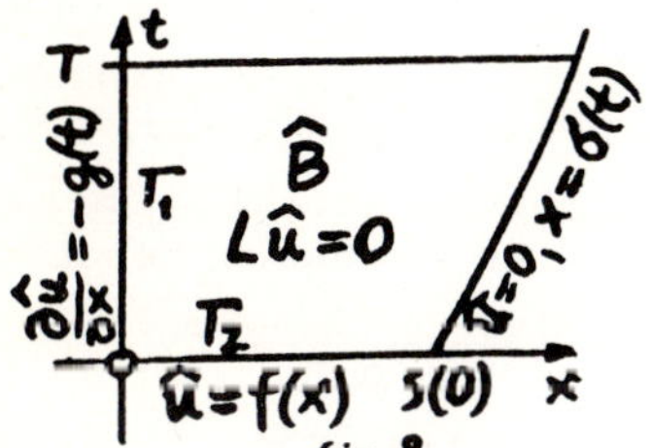

fig.8

$\hat{B}=\{0<t<T,\quad 0<x<\sigma(t)$

with the boundary condition

$$(5.6)\begin{cases}\frac{\partial \hat{u}}{\partial x}=-g(t) \text{ on } \Gamma_1,\ \hat{u}=f(x) \text{ on } \Gamma_2,\\ \hat{u}=0 \text{ on } x=\sigma(t);\end{cases}$$

here $\sigma(t)$ is an arbitrarily chosen function of t of the class C^1, defined in $[0,T]$ with $\sigma(0)=s(0)$, $\frac{d\sigma}{dt}\geq 0$. The integral equation is solved by iteration, starting with $s_o(t)\in C^1$ and determining $s_{n+1}(t)$ from $s_n(t)$ $(n=0,1,2,\dots)$ by

$$(5.7)\quad s_{n+1}(T) = \Phi(s_n(T)) \text{ with } s_{n+1}(0)=s(0).$$

Now $\hat{u}$ depends on $\sigma(t)$ monotonically non-decreasing by classical monotonicity properties for the heat conduction equation; therefore $\Phi(s)$ is an antitone operator in s(t) (Hoffmann [77]) and one can apply the fixed point theorem of Schauder and the theory for iteration procedures for antitone operators (compare f.i. Collatz [66] p. 352).
If one starts with two functions $v_o(t)$ and $w_o(t)\in C^1$ and determines

$$v_1(t) = \Phi(w_o(t)),\quad w_1(t) = \Phi(v_o(t)),$$

and if (5.8) $v_o(t) \leq v_1(t) \leq w_1(t) \leq w_o(t)$ for $t \in [0,T]$,

then one has the existence of a solution s(t) of (5.5) with the error bound

(5.9) $\quad v_1(t) \le s(t) \le w_1(t)$.

Another kind of monotonicity in Stefan problems was considered by Glashoff-Werner [77] .

<u>Numerical example:</u> $s(0) = 0, \quad g(t) = 1.$

We can choose a solution $\hat{u}$ of $L\hat{u}=o$ which satisfies the boundary conditions (here Γ_2 is simply the origin) for arbitrary real a,b

$$\hat{u}=-x+a\left(\frac{x^2}{2}+t\right) + b\left(\frac{x^4}{24}+\frac{x^2 t}{2} + \frac{t^2}{2}\right).$$

For simplicity we take b=0; then $\hat{u}=o$ gives

$$t = \frac{x}{a} - \frac{x^2}{2} \quad \text{or } x = s_o(t)$$

The iteration (5.7) is

$$s_1(t)=t-\int_0^{s_o(t)} \hat{u}dx=t+\frac{s_o{}^2(t)}{2}-\frac{a}{6}\left[s_o(t)\right]^3-at\ s_o(t).$$

$$\text{or} \quad s_o(t) = \frac{1}{a}\left(1-\sqrt{1-2a^2t}\ \right)$$

$$s_1(t) = \frac{1}{3a^2}\left(1-\sqrt{1-2a^2t}^{\ 3}\right);$$

fig. 9 shows the graphs of the functions $s_o(t)$ and $s_1(t)$ for a=1 and $a=\frac{1}{2}$.

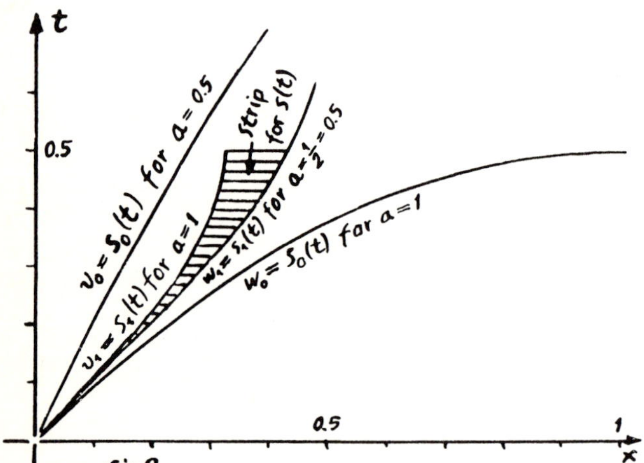

fig.9.

If we choose $s_o(t)$ for $a=\frac{1}{2}$ as v_o and $s_o(t)$ for a=1 as w_o, then (5.8) is satisfied, fig.9., for $0<t\le\frac{1}{2}$ and the error bound (5.9) holds.

We can improve the error bounds ; f.i. for

$t=\frac{1}{2}$	a	b	s_0	s_1
	1	$-\frac{48}{73}$	$\frac{1}{2}$	0.4023
	0.6925	0	0.4023	0.4341

$$\left| 0.4182 - s\left(\tfrac{1}{2}\right)\right| \le 0.0159.$$

I thank Mr. Uwe Grothkopf for numerical calculations on a computer.

References

Baiocchi, C., [74] Free boundary problems in the theory of fluid flow through porous media. Proc. Intern. Congress of Math. 1974 Vancouver, Vol. 2

Bredendiek, E., [7o] Charakterisierung und Eindeutigkeit bei simultanen Approximationen, Z.angew. Math. Mech. 5o (197o), 4o3-41o.

Bredendiek, E.- L. Collatz [76] Simultan-Approximation bei Randwertaufgaben, Intern. Ser.Num. Math. 3o (1976) 147-174.

Collatz, L.,[66] Functional Analysis and Numerical Mathematics Acad. Press 1966, 473 p.

Collatz, L.,[7o] Einseitige Tschebyscheff-Approximation bei Randwertaufgaben. Proc. Internat. Conference on Constructive funct. Theory, Varna, Bulgaria 197o, 151-162

Collatz, L.,[73] Methods for Solution of Partial Differential Equations Symp. Numer. Solution Part. Diff. Equ. Kjeller, Norway 1973, 1-16

Glashoff, K., - B. Werner [77] Monotonicity in Stefan Problem Conference Num. Treatment Diff. Equ. Oberwolfach May 1977, to appear in Internat. Ser. Num. Math.

Evans, G.W. [51] A note on the existence of a solution of a Stefan problem

Hoffmann, K.H., [77] Lecture on free boundary value problems, Conference Num. Solut. Diff. Equ. Oberwolfach May 1977 to appear in Internat. Ser. Num. Math.

Mitchell, A.R. - R. Wait [77] The finite Element Method in Partial Differential Equations, 1977, 192 p.

Wetterling, W., [76] Lecture on error bounds for solutions of singular elliptic boundary value problems at Mathem. Forschungsinstitut Oberwolfach, to appear in Internat. Ser. Num. Math.

Whiteman, J.R. [73], The mathematics of Finite Elements and applications, Acad. Press 1973, 52o S.

Whiteman, J.R., [76] Lecture about finite element methods for singular elliptic boundary value problems at Mathem. Forschungs-Institut Oberwolfach, 19. May 1976, to appear in Internat. Ser.Num.Math.

Whiteman, J.R. - N. Papamichael [72] Treatment of Harmonic Mixed Boundary Problems by Conformal Transformation methods Journ.Appl. Math. Phys. (ZAMP) 23 (1972) 655-664

Whiteman, J.R. and B. Schiff [76], Finite element approximation of singular functions, Internat. Series Numer. Math. 3o (1976) 319-333.

THE INCORPORATION OF BOUNDARY CONDITIONS IN SPLINE APPROXIMATION PROBLEMS

M G Cox

Abstract

The problem of determining polynomial spline approximations subject to derivative boundary conditions is considered. It is shown that if B-splines are employed as a basis, the incorporation of a variety of boundary conditions and the solution of the resulting equations for the B-spline coefficients can be carried out in an efficient and stable manner. The methods discussed can be applied to interpolating splines and to splines which approximate discrete data sets in the least squares sense. The boundary conditions can normally be incorporated by extending and refining existing algorithms rather than by designing entirely new ones.

1 Introduction

We consider the numerical solution of approximation problems, using polynomial splines, in which it is required that the spline $s(x)$ and/or some of its derivatives take prescribed values at the boundaries of a given interval. Common cases are (i) spline interpolation in which $s(x)$ is to assume given derivative end conditions and (ii) spline fitting to discrete data sets in which $s(x)$ is to satisfy prescribed end conditions exactly, but to approximate the bulk of the data in the least squares sense.

Methods for forming and incorporating such constraints are discussed in the case where $s(x)$ is expressed in terms of B-splines, a representation recognised as possessing significant advantages in terms of flexibility, efficiency and stability.

2 B-spline representation

Let N and n be prescribed positive integers and $[a, b]$ a finite interval of the real line. Let $\pi = \{\lambda_j\}_0^N$ be an n-extended partition of $[a, b]$, viz.

$$a = \lambda_0 < \lambda_1 \leq \lambda_2 \leq \ldots \leq \lambda_{N-1} < \lambda_N = b \tag{2.1}$$

with

$$\lambda_j < \lambda_{j+n} \quad (j = 0, 1, \ldots, N-n). \tag{2.2}$$

For $j = 1, 2, \ldots, N + n - 1$, denote by $N_{nj}(x)$ the normalized B-spline of order n (degree $n - 1$) defined upon the knots $\{\lambda_k\}_{k=j-n}^{j}$, i.e.

$$N_{nj}(x) = (\lambda_j - \lambda_{j-n})M_{nj}(x),$$

where $M_{nj}(x)$ is the nth divided difference at $\lambda=\lambda_{j-n}, \lambda_{j-n+1}, \ldots, \lambda_j$ of the truncated power function

$$(\lambda-x)_+^{n-1} = \{\max(\lambda-x, 0)\}^{n-1} .$$

Then $\{N_{nj}(x)\}_{j=1}^{N+n-1}$ forms a basis for polynomial splines of order n on π in $[a, b]$ (Curry and Schoenberg (1966)). Here, in order to define the complete set of B-splines required, additional knots satisfying

$$\lambda_{1-n} \leq \lambda_{2-n} \leq \ldots \leq \lambda_{-1} \leq a; \; b \leq \lambda_{N+1} \leq \lambda_{N+2} \leq \ldots \leq \lambda_{N+n-1} \qquad (2.3)$$

are introduced.

Methods exist for evaluating the B-spline basis (de Boor (1972), Cox (1972)) and the B-spline representation

$$s(x) = \sum_{j=1}^{N+n-1} c_j N_{nj}(x) \quad (a \leq x \leq b) \qquad (2.4)$$

(de Boor (1972)), and their unconditional numerical stability has been established (Cox (1972), Cox (1976)). In (2.4), s(x) is an arbitrary polynomial spline of order n on π and $\{c_j\}_{j=1}^{N+n-1}$ are the (unique) B-spline coefficients in the B-spline representation of s(x). In order to determine values for the coefficients c_j in (2.4), N+n-1 independent pieces of information relating to the values of s(x) and its derivatives are normally provided.

The following properties (de Boor (1972)) of B-splines will be employed.

Non-negativity: $N_{nj}(x) \geq 0.$ (2.5)

Normalization: $\sum_j N_{nj}(x) \equiv 1.$ (2.6)

Compact support: $N_{nj}(x) \equiv 0$ outside $[\lambda_{j-n}, \lambda_j]$. (2.7)

3 The choice of additional knots

There are many possible choices for the additional knots in (2.3). One class of choices is given by

$$\lambda_j = \begin{cases} \lambda_0 + jd_1 & (j < 0) \\ \lambda_N + (N-j)d_2 & (j > N), \end{cases}$$

with $d_1, d_2 \geq 0$, three members of this class being (i) $d_1 = \lambda_1 - \lambda_0$, $d_2 = \lambda_N - \lambda_{N-1}$, (ii) $d_1 = d_2 = (\lambda_N - \lambda_0)/N$, and (iii) $d_1 = d_2 = 0$.

In the special case of uniformly spaced knots, (i) and (ii) have the desirable property that all the resulting B-splines are simple translations (shifts) of any one of them. On the other hand, choice (iii), giving coincident knots of multiplicity n at each boundary, has the following two advantages with all knot sets. First, it enables a wide variety of boundary conditions to be incorporated in an efficient and stable manner (see later sections). Second, the condition numbers of the resulting matrices arising in interpolation and least squares problems appear to be smaller. In an empirical comparison (Cox (1975b)) of the spectral condition numbers of the matrices associated with a range of some twenty practical cubic spline approximation problems, values between (i) 23 and 65, (ii) 16 and 170 and (iii) 4.5 and 7.8 were observed, respectively, for the above three choices. In fact, Kozak (1976) has established that for a class of cubic spline interpolation problems the condition number with respect to the maximum norm is minimal for the choice (iii).

Thus, while it is sometimes desirable to construct special algorithms to take advantage of particular circumstances, in the case of general-purpose algorithms for incorporating boundary conditions we prefer to work with coincident knots at the boundaries $x = a$ and $x = b$.

4 The evaluation of B-splines and their derivatives

In setting up systems of equations defining the B-spline coefficients it is necessary to compute values of the B-splines and their derivatives. These values can be formed from the fundamental recurrence relations

$$N^{(r)}_{nj}(x) = \left[\frac{n-1}{n-r-1}\right]\left\{\left[\frac{x-\lambda_{j-n}}{\lambda_{j-1}-\lambda_{j-n}}\right]N^{(r)}_{n-1,j-1}(x)+\left[\frac{\lambda_j-x}{\lambda_j-\lambda_{j-n+1}}\right]N^{(r)}_{n-1,j}(x)\right\} \qquad (4.1)$$

and

$$N^{(r)}_{nj}(x) = (n-1)\left[\frac{N^{(r-1)}_{n-1,j-1}(x)}{\lambda_{j-1}-\lambda_{j-n}} - \frac{N^{(r-1)}_{n-1,j}(x)}{\lambda_j-\lambda_{j-n+1}}\right], \qquad (4.2)$$

with the starting conditions

$$N_{1j}(x) = \begin{cases} 1 & (\lambda_{j-1} \leq x < \lambda_j) \\ 0 & (\text{otherwise}) . \end{cases} \qquad (4.3)$$

(The strict inequality in (4.3) is permitted to include equality in the case $j = N$, so that $x = b$ is contained in the right-most interval.) For derivations of (4.1) in the case $r = 0$ see Cox (1972) and de Boor (1972) – the latter author also establishes (4.2); proofs of (4.1) for $r > 0$ are given in Cox (1975b) and Butterfield (1976).

Note that (4.1) and (4.3) and the fact that $\lambda_0 = a$ and $\lambda_N = b$ imply

$$N_{nj}(a) = N_{n,N+n-j}(b) = \begin{cases} 1 & (j = 1) \\ 0 & (j > 1) \end{cases} . \tag{4.4}$$

The following property of B-spline derivatives is established in Cox (1977):

$$\text{sign}\,\{N_{nj}^{(r)}(a)\} = \begin{cases} (-1)^{j-r+1} & (j \leq r+1) \\ 0 & (j > r+1) \end{cases} \tag{4.5a}$$

and

$$\text{sign}\,\{N_{nj}^{(r)}(b)\} = \begin{cases} (-1)^{N+n-1-j} & (j \geq N+n-1-r) \\ 0 & (j < N+n-1-r) . \end{cases} \tag{4.5b}$$

The algorithm below, based on this property and relation (4.2), evaluates the non-zero derivatives of order r of the B-spline basis of order n, at $x = a$ or $x = b$. Such values are required in forming equations (7.1a) and (7.1c), for example.

> Set $N_{n-r,1} = 1$.
> For $k = 1, 2, \ldots, r$
>   for $j = 1, 2, \ldots, k+1$
>   form $N_{n-r+k,j}^{(k)}$ using (4.2).

All values required by this algorithm which are undefined are to be taken as zero. The algorithm forms successive columns in a triangular array exemplified here by the case $n = 6$, $r = 3$:

$$\begin{array}{cccc} & & & N'''_{61} \\ & & N''_{51} & \\ & N'_{41} & & N'''_{62} \\ N_{31} & & N''_{52} & \\ & N'_{42} & & N'''_{63} \\ & & N''_{53} & \\ & & & N'''_{64} \end{array} . \tag{4.6}$$

As a result of (4.5), the elements in (4.6) have the sign pattern

$$
\begin{array}{cccccccccc}
 & & & - & \qquad\qquad & & & & + \\
 & & + & & & & & + & \\
 & - & & + & & & + & & - \\
+ & & - & & & + & & - & \\
 & + & & - & & & - & & + \\
 & & + & & & & & + & \\
 & & & + & & & & & - \\
 & & & & & & & & \\
 & (x = a) & & & & & (x = b) & &
\end{array}
$$

from which it is apparent that, in using (4.2), differences between numbers of like sign are never formed. It follows that no loss of precision due to cancellation can occur. In fact, the computed value $\bar{N}_{nj}^{(r)}$ of $N_{nj}^{(r)}$ has a very small relative error; the _a priori_ bound

$$| \bar{N}_{nj}^{(r)} - N_{nj}^{(r)} | \leqslant 4.9r2^{-t} |N_{nj}^{(r)}| \quad , \tag{4.7}$$

where t is the number of binary digits in the mantissa of the floating point word, is established in Cox (1977).

5 Derivative-free spline interpolation

We outline briefly some existing methods for spline interpolation without constraints; our algorithms which allow boundary conditions are essentially extensions of these methods.

Suppose $m = N + n - 1$ values of a function $f(x)$ are given at a strictly increasing set of values $\{x_i\}_{i=1}^{m}$, lying within an interval (a,b) of an independent variable x, and knots $\{\lambda_j\}_{j=0}^{N}$ satisfying (2.1) and (2.2) are provided. A unique solution to the _spline interpolation problem_

$$s(x_i) = f(x_i) \quad (i = 1, 2, \ldots, m) \tag{5.1}$$

exists if and only if the inequalities

$$x_j < \lambda_j < x_{j+n} \quad (j = 1, 2, \ldots, N - 1) \tag{5.2}$$

are satisfied (Schoenberg and Whitney (1953)). If the conditions (5.2) are assumed to hold, the values of c_j may be determined by solving the system of m equations

$$\sum_{j=1}^{m} c_j N_{nj}(x_i) = f(x_i) \quad (i = 1, 2, \ldots, m) \ . \tag{5.3}$$

Details of an algorithm for forming and solving this system are given in Cox (1975a). Full advantage is taken of the compact support property (2.7) of the B-splines, with the result that the required coefficients can be computed in a total number of arithmetic operations proportional to mn^2. Note that the use of a basis not having the compact support property may result in an operation count proportional to m^3. In Cox (1975a) it was proposed that the system (5.3) which, because of (2.7), has a coefficient matrix that is stepped banded of bandwidth n*, be solved in a stable manner using a variant of Gaussian elimination with partial pivoting. It was also stated that at the expense of a greater operation count, although still proportional to mn^2, the system could be solved, without the need for pivoting, using orthogonal transformations, viz by applying a special sequence of Householder transformations (Reid (1967)) or by employing classical or square-root-free plane rotations in the manner suggested by Gentleman (1973). Both these approaches avoid the theoretical possibility of growth associated with elimination methods. Recently, however, de Boor and Pinkus (1977) have proved that the total positivity[‡] of the system matrix, a property established by Karlin (1968), is sufficient to ensure that (5.3) can be solved safely by Gaussian elimination <u>without pivoting</u>.

6 Spline interpolation with boundary conditions

A generalization, incorporating boundary conditions, of the spline interpolation problem of Section 5 is as follows. Given m data points $(x_i, f(x_i))$ $(i = 1, 2, \ldots, m)$, with $a \leqslant x_1 < x_2 < \ldots < x_m \leqslant b$, and a total of $p + q$ independent boundary conditions, p $(0 \leqslant p \leqslant n)$ at $x = a$ and q $(0 \leqslant q \leqslant n)$ at $x = b$, determine a spline $s(x)$ of order n passing through these points which satisfies the specified boundary conditions. Note that if $p > 0$, we require $a < x_1$ and if $q > 0$, $b > x_m$.

It will be assumed that knots $\{\lambda_j\}_{j=0}^{N}$, where $N = m - n + p + q + 1$, satisfying (2.1) and (2.2) are provided. Normally, but not necessarily, n is even and $p = q = n/2$, in which case $\lambda_j = x_j$ $(j = 1, 2, \ldots, m)$ is a possible choice. Our treatment can readily be specialized if required to such circumstances to yield slightly more efficient algorithms. It will also be assumed that the following generalization of the Schoenberg-Whitney conditions (5.2) holds:

$$u_j < \lambda_j < u_{j+n} \quad (j = 1, 2, \ldots, N - 1) \quad , \tag{6.1}$$

---

*Such a matrix has the property that each row has at most n non-zero elements, which occur in adjacent positions, and that the number of the column containing the first non-zero element is a non-decreasing functions of row number.

‡A totally positive matrix has all its minors non-negative.

where

$$(u_1, u_2, \ldots, u_{N+n-1}) \equiv (\underbrace{a, a, \ldots, a}_{p}, x_1, x_2, \ldots, x_m, \underbrace{b, b, \ldots, b}_{q}).$$

The choice of the above value of N and the satisfaction of these conditions ensures a unique solution.

It will further be assumed that the boundary conditions can be expressed in the form

$$s^{(r)}(a) = f^{(r)}(a) \quad (r = k_1, k_2, \ldots, k_p)$$

and

$$s^{(r)}(b) = f^{(r)}(b) \quad (r = \ell_1, \ell_2, \ldots, \ell_q),$$

where $0 \leq k_1 < k_2 < \ldots < k_p < n$ and $0 \leq \ell_1 < \ell_2 < \ldots < \ell_q < n$.

Three methods, which we term the <u>direct</u>, <u>data modification</u> and <u>basis modification</u> methods for imposing such constraints are discussed; each retains the stepped banded structure of the unconstrained method. The first method is completely general and consists of appending in a natural way to equations (5.1) those involving the boundary information, giving a total of $m + p + q = N + n - 1$ equations to be solved. The second method is applicable to the commonly occurring case in which the prescribed boundary conditions are the values of $f^{(r)}(a)$ $(r = 0, 1, \ldots, p - 1)$ and $f^{(r)}(b)$ $(r = 0, 1, \ldots, q - 1)$. It involves the initial determination of $p + q$ of the B-spline coefficients which are strongly related to the boundary conditions, a simple modification to the data, and the subsequent solution of a system of m equations. The third method, which can be applied particularly easily when single boundary conditions are given, consists of replacing the B-spline basis by a modified basis which satisfies the boundary conditions, followed by the solution of m equations for the coefficients of the modified basis. The last two methods are somewhat more restrictive than the first, but have distinct advantages in certain generalizations of the interpolation problem to least squares data fitting by splines (see Section 10, for example). Our treatment of the second and third methods will be given only for boundary conditions at $x = a$ since those at $x = b$ may be treated similarly.

7 The direct method of imposing derivative boundary conditions

The system of $N + n - 1$ equations defining the B-spline coefficients in the solution to the problem outlined in Section 6 may be expressed as

$$\sum_{j=1}^{r+1} c_j N_{nj}^{(r)}(a) = f^{(r)}(a) \qquad (r = k_1, k_2, \ldots, k_p), \tag{7.1a}$$

$$\sum_{j=1}^{N+n-1} c_j N_{nj}(x_i) = f(x_i) \qquad (i = 1, 2, \ldots, m) , \tag{7.1b}$$

$$\sum_{j=N+n-1-r}^{N+n-1} c_j N_{nj}^{(r)}(b) = f^{(r)}(b) \qquad (r = \ell_1, \ell_2, \ldots, \ell_q) , \tag{7.1c}$$

the limits on the summation symbols in (7.1a) and (7.1c) resulting from property (4.5).

All the elements of the matrix of coefficients of (7.1) are values of B-splines or their derivatives and may be formed using the recurrence relations of Section 4.

As a result of the ordering of the equations, system (7.1) has a stepped banded system matrix of bandwidth n. If there is at least one derivative boundary condition this matrix is not totally positive, there being at least one negative element as a result of (4.5), so a method such as Gaussian elimination with partial pivoting should be used for the solution of the system.

The approach outlined here has been employed by Herriot (1976) for natural spline interpolation. In this case the spline is of even order $n = 2n'$ and the derivative boundary conditions take the form

$$\sum_{j=1}^{r+1} c_j N_{nj}^{(r)}(a) = \sum_{j=N+n-1-r}^{N+n-1} c_j N_{nj}^{(r)}(b) = 0 \qquad (r = n', n'+1, \ldots, 2n'-2).$$

A word about scaling is in order. The 1-norms of the rows of the matrix corresponding to the equations (7.1b) are all equal to unity, as a consequence of (2.5) and (2.6). In order to equilibrate all rows of the matrix, it is recommended that the remaining equations be scaled to have this property; this has the additional advantage that the equations become invariant with respect to linear transformations, and hence to the units of measurement, of the independent variable x.

8 The imposition of boundary conditions by data modification

If the left hand boundary conditions correspond to the specification of the value and the first p-1 derivatives of $s(x)$ at $x = a$ then, by virtue of (7.1), the first p B-spline coefficients are defined by the triangular system of equations

$$\sum_{j=1}^{r+1} c_j N_{nj}^{(r)}(a) = f^{(r)}(a) \qquad (r = 0, 1, \ldots, p-1) . \tag{8.1}$$

Because of (4.7) the values of the B-spline derivatives required in (8.1) may be computed in an unconditionally stable manner. The system is readily solved by the usual process of forward substitution.

Once the values of $c_1, c_2, \ldots, c_p$ defined by (8.1) have been obtained we consider the modified spline

$$\tilde{s}(x) = s(x) - \sum_{j=1}^{p} c_j N_{nj}(x) = \sum_{j=p+1}^{N+n-1} c_j N_{nj}(x) .$$

Then, accordingly, modified function values $f(x_i)$ given by

$$\tilde{f}(x_i) = f(x_i) - \sum_{j=1}^{p} c_j N_{nj}(x_i) \quad (i = 1, 2, \ldots, m)$$

are introduced and the modified data approximated by $\tilde{s}(x)$. Note that only data in the interval $a \leq x < \lambda_p$ has to be so modified since for $x \geq \lambda_p$, $N_{nj}(x) \equiv 0$ $(j = 1, 2, \ldots, p)$. Boundary conditions at $x = b$ are treated in a similar fashion. (Cf Clenshaw and Hayes (1965) for a similar approach employing polynomials as the approximating functions. In the case of polynomials (and in fact of any function which cannot be represented as a linear combination of basis functions with compact support), however, all data has of course to be modified appropriately.)

Note that the resulting matrix of coefficients is still stepped banded, it being similar to that of an unconstrained problem, but with the first p and the last q columns deleted.

9 The imposition of boundary conditions by basis modification

The basic idea, in which the B-spline basis is replaced by a modified basis, can best be illustrated by a simple example.

Let r $(0 < r < n)$ and the value of $f^{(r)}(a)$ be prescribed. The constraint

$$s^{(r)}(a) = f^{(r)}(a) \tag{9.1}$$

can be enforced as follows. First, suppose $r = 1$. Then from (7.1),

$$c_1 N_{n1}'(a) + c_2 N_{n2}'(a) = f'(a) . \tag{9.2}$$

But, because of (4.5), $N_{n1}'(a) \neq 0$. Hence, the elimination of c_1 from (2.4) by means of (9.2) gives

$$s(x) - \{f'(a)/N_{n1}'(a)\} N_{n1}(x) = \sum_{j=2}^{N+n-1} c_j \tilde{N}_{nj}(x) = \tilde{s}(x) , \tag{9.3}$$

say, where

$$\tilde{N}_{nj}(x) = \begin{cases} N_{n2}(x) - \{N_{n2}'(a)/N_{n1}'(a)\} N_{n1}(x) & (j = 2) \\ N_{nj}(x) & (j > 2) . \end{cases}$$

But properties (2.6) and (2.7) yield $N_{n1}'(a) + N_{n2}'(a) = 0$. Thus

$$\tilde{N}_{nj}(x) = \begin{cases} N_{n1}(x) + N_{n2}(x) & (j = 2) \\ N_{nj}(x) & (j > 2) . \end{cases}$$

Consequently, if the appropriate values of the expression $\{f'(a)/N_{n1}'(a)\} N_{n1}(x)$ are subtracted from the data (only the values of $x < \lambda_1$ are affected), the use of the modified representation (9.3) enables, in the case $r = 1$, the condition (9.1) to be incorporated automatically. Note that the function $\tilde{N}_{n2}(x)$ has the same support as $N_{n2}(x)$, is non-negative and, moreover, can be formed stably since it is the sum of two non-negative quantities each having a small relative error.

In the natural generalization of this approach to any value of r $(0 < r < n)$, we obtain

$$\tilde{s}(x) = s(x) - \{f^{(r)}(a)/N_{n1}^{(r)}(a)\} N_{n1}(x) = \sum_{j=2}^{N+n-1} c_j \tilde{N}_{nj}(x) , \qquad (9.4)$$

where

$$\tilde{N}_{nj}(x) = \begin{cases} N_{nj}(x) - \{N_{nj}^{(r)}(a)/N_{n1}^{(r)}(a)\} N_{n1}(x) & (j \leqslant r+1) \\ N_{nj}(x) & (j > r+1) . \end{cases} \qquad (9.5)$$

Unfortunately, no longer do all the modified basis functions have the property that they are formed as positive linear combinations of non-negative quantities, and hence there is no guarantee that they can be computed with small relative errors, although their absolute errors will certainly be small compared with unity, the maximum possible value of $N_{nj}(x)$.

In order to obtain basis functions which can be formed with small relative errors, consider again the representation (9.4), but with $\tilde{N}_{nj}(x)$ now defined by

$$\tilde{N}_{nj}(x) = \begin{cases} N_{nj}(x) - \{N_{nj}^{(r)}(a)/N_{n,j-1}^{(r)}(a)\} N_{n,j-1}(x) & (j \leqslant r + 1) \\ N_{nj}(x) & (j > r + 1) . \end{cases} \qquad (9.6)$$

As with the representation (9.4) and (9.5) it is readily verified that $\tilde{s}^{(r)}(a) = 0$ and $s^{(r)}(a) = f^{(r)}(a)$, as required. Moreover, both representations enjoy the property that for $j = 2, 3, \ldots, N+n-1$ the basis functions $\tilde{N}_{nj}(x)$ have the same support as the $N_{nj}(x)$. However, the representation (9.4) and (9.6) has the distinct advantage that the factors $N_{nj}^{(r)}(a)/N_{n,j-1}^{(r)}(a)$ are all negative by virtue of (4.5), and hence that the $\tilde{N}_{nj}(x)$ are formed as positive linear combinations of non-negative quantities, with the consequence that the computed values have small relative errors.

Again, equilibration is desirable following the formation of the system of equations defining the coefficients of the modified basis.

The ideas of this section can be extended to more complicated situations. For instance, the basis modification approach is proposed by Greville (1969) for computing natural interpolating splines and by Hayes and Halliday (1974) for the imposition of boundary constraints in fitting cubic spline surfaces by least squares.

10 Least squares spline approximation with boundary conditions

The problem of least squares spline approximation with boundary conditions can be posed as the following natural extension of the spline interpolation problem of Section 6. Suppose m data points $(x_i, f(x_i))$ $(i = 1, 2, \ldots, m)$, with $a \leq x_1 \leq x_2 \leq \ldots \leq x_m \leq b$, and p+q boundary conditions, as in Section 6 (with $a < x_1$ if $p > 0$ and $b > x_m$ if $q > 0$), are provided. Note that equalities are now permitted among the x-values, corresponding, in an experimental situation, to the replication of measurements. The problem is to determine a spline s(x) of order n which matches the specified boundary conditions and which satisfies the remainder of the data in the least squares sense. It will be assumed that $\{\lambda_j\}_{j=0}^{N}$ where $N \leq \tilde{m} - n + p + q + 1$ and $\tilde{m}$ is the number of distinct values of x_i, satisfying (2.1) and (2.2), are provided. The B-spline coefficients c_j are then given by the least squares solution of the over-determined set of linear equations

$$\sum_{j=1}^{N+n-1} c_j N_{n,j}(x_i) \approx f(x_i) \quad (i = 1, 2, \ldots, m) , \tag{10.1}$$

subject to the constraints (7.1a) and (7.1c). Weighted least squares solutions may be obtained by first multiplying each equation in (10.1) by an appropriate weight.

The solution to this problem is unique if and only if there is at least one ordered subset (strictly ordered with respect to the x-values)

$$(u_1, u_2, \ldots, u_{N+n-1}) \in (\underbrace{a, a, \ldots, a}_{p}, x_1, x_2, \ldots, x_m, \underbrace{b, b, \ldots, b}_{q}).$$

which satisfies the Schoenberg-Whitney conditions (6.1).

If there are no constraints present (i.e. $p = q = 0$), the orthogonalization methods due to Reid (1967) and Gentleman (1973) referred to in Section 5 are strong candidates for the method of solution since the former takes advantage of structure and the latter can readily be organised to do so. In the presence of constraints the problem may be solved using one of the methods employing Householder transformations given by e.g. Hayes and Halliday (1974) or Lawson and Hanson (1974) for linear least squares problems with general linear equality constraints. Unfortunately, these methods destroy the structure of the matrices associated with the problem, because of the need to perform column interchanges to ensure stability.

A stable method based upon the use of Givens rotations which preserves most of the structure is proposed in Cox (1975b). However, for the important class of problems in which the boundary conditions consist of the values of $s^{(r)}(a)$ ($r = 0, 1, \ldots, p-1$) and $s^{(r)}(b)$ ($r = 0, 1, \ldots, q-1$), the method of data modification (Section 8) enables the problem to be converted into one which is unconstrained and in which all structure can be preserved.

The method of basis modification (Section 9) also lends itself readily to least squares spline approximation with boundary conditions.

Acknowledgement

Mr E L Albasiny and Mr J G Hayes made many valuable comments on the first draft of this paper.

References

BOOR, C. DE. On calculating with B-splines. J. Approximation Theory, 1972, 6, 50-62.

BOOR, C. DE and PINKUS, A. Backward error analysis for totally positive linear systems. Numer. Math., 1977, 27, 485-490.

BUTTERFIELD, K.R. The computation of all the derivatives of a B-spline basis. J. Inst. Math. Appl., 1976, 17, 15-25.

CLENSHAW, C.W. and HAYES, J.G. Curve and surface fitting. J. Inst. Math. Appl., 1965, 1, 164-183.

COX, M.G. The numerical evaluation of B-splines. J. Inst. Math. Appl., 1972, 10, 134-149.

COX, M.G. An algorithm for spline interpolation. J. Inst. Math. Appl., 1975a, 15, 95-108.

COX, M.G. Numerical methods for the interpolation and approximation of data by spline functions. London, City University, PhD Thesis, 1975b.

COX, M.G. The numerical evaluation of a spline from its B-spline representation. National Physical Laboratory NAC Report No. 68, 1976. To appear in J. Inst. Math. Appl.

COX, M.G. The incorporation of boundary conditions in spline approximation problems. National Physical Laboratory NAC Report No. 80, 1977.

CURRY, H.B. and SCHOENBERG, I.J. On Pólya frequency functions IV: the fundamental spline functions and their limits. J. Analyse Math., 1966, 17, 71-107.

GENTLEMAN, W.M. Least squares computations by Givens transformations without square roots. J. Inst. Math. Appl., 1973, 12, 329-336.

GREVILLE, T.N.E. Introduction to spline functions. Theory and Application of Spline Functions, edited by T.N.E. Greville, New York, Academic Press, 1969, 1-35.

HAYES, J.G. and HALLIDAY, J. The least-squares fitting of cubic spline surfaces to general data sets. J. Inst. Math. Appl., 1974, 14, 89-103.

HERRIOT, J.G. Calculation of interpolating natural spline functions using de Boor's package for calculating with B-splines. Stanford University CS Report No. 569, 1976.

KARLIN, S. Total positivity Vol. I, Stanford, Stanford University Press, 1968.

KOZAK, J. Private communication, 1976.

LAWSON, C.L. and HANSON, R.J. Solving least squares problems. Englewood Cliffs, New Jersey, Prentice-Hall, 1974.

REID, J.K. A note on the least squares solution of a band system of linear equations by Householder reductions. Comput. J., 1967, 10, 188-189.

SCHOENBERG, I.J. and WHITNEY, Anne. On Pólya frequency functions III. Trans. Am. Math. Soc., 1953, 74, 246-259.

A Time-Stepping Method for Galerkin Approximations for Nonlinear Parabolic Equations

Jim Douglas, Jr.*, Todd Dupont*, and Peter Percell**

Abstract. A modified backward difference time discretization is considered for Galerkin approximations to the solution of the nonlinear parabolic equation $c(x,u)u_t - \nabla\cdot(a(x,u)\nabla u) = f(x,u)$. This procedure allows efficient use of such direct methods for solving linear algebraic equations as nested dissection. Optimal order error estimates and almost optimal order work requirements are derived.

1. Introduction. We shall consider the numerical solution of the nonlinear parabolic problem

$$
\text{(1.1)}\quad
\begin{aligned}
&\text{(a)}\quad c(x,u)\frac{\partial u}{\partial t} - \nabla\cdot(a(x,u)\nabla u) = f(x,u), && x\in\Omega,\ t\in J,\\
&\text{(b)}\quad \frac{\partial u}{\partial n} = g(x,t), && x\in\partial\Omega,\ t\in J,\\
&\text{(c)}\quad u = u_0(x), && x\in\Omega,\ t=0,
\end{aligned}
$$

where Ω is a bounded domain in $\mathbb{R}^d$, $d = 2$ or 3, $\partial\Omega$ is smooth and $J = (0,T]$, by a Galerkin approximation in the space variables and a new discretization in the time variable that is intended to offer a very significant reduction in the computing requirements to evaluate the approximate solution. Denote the $L^2(\Omega)$-inner product by (v,w) and the $L^2(\partial\Omega)$-inner product by $\langle v,w\rangle$, respectively. Let $W_p^j(\Omega)$ be the Sobolev space of functions $v\in L^p(\Omega)$ such that $D^\alpha v\in L^p(\Omega)$, $|\alpha|\le j$, for $1\le p\le\infty$, and norm $W_p^j(\Omega)$ in the usual manner. We shall let $\|v\|$ indicate the

---

*University of Chicago, Chicago, Illinois 60637, U.S.A.

**University of Houston, Houston, Texas 77004, U.S.A.

$L^2(\Omega)$-norm and $\|v\|_j$ the norm in $W_2^j(\Omega) = H^j(\Omega)$.

For $0 < h < 1$, let $\mathcal{M}_h \subset W_\infty^1(\Omega)$ satisfy the following constraints:

(a) For $z \in W_p^k(\Omega)$, where $1 \leq p \leq \infty$ and $1 \leq k \leq r$,

$$\inf_{\chi \in \mathcal{M}_h} \|z - \chi\|_{W_p^j(\Omega)} \leq C \|z\|_{W_p^k(\Omega)} h^{k-j}, \quad j = 0 \text{ or } 1,$$

(1.2) (b) For $\chi \in \mathcal{M}_h$, $\|\chi\|_{W_\infty^j(\Omega)} \leq Ch^{-d/2} \|\chi\|_j$, $j = 0$ or 1,

(c) For $\chi \in \mathcal{M}_h$, $\|\chi\|_1 \leq Ch^{-1} \|\chi\|$.

A well-known Galerkin method [3, 8] for (1.1) is given by the use of backwards differencing in time combined with lagging the evaluation of the coefficients. Let $u_h: \{0 = t_0, \Delta t = t_1, \ldots, t_N = T\} \rightarrow \mathcal{M}_h$ be determined by the relations

(1.3)

(a)
$$\begin{cases} (a(u_0)\nabla(u_h^0 - u_0), \nabla\chi) = 0, & \chi \in \mathcal{M}_h, \\ (u_h^0 - u_0, 1) = 0, & \end{cases}$$

(b)
$$(c(u_h^{n-1})\partial u_h^n, \chi) + (a(u_h^{n-1})\nabla u_h^n, \nabla\chi) = (f(u_h^{n-1}), \chi) + \langle g^n, \chi \rangle, \quad \chi \in \mathcal{M}_h,$$

where the writing of the x and t variables has been suppressed, $\partial u_h^n = (u_h^n - u_h^{n-1})/\Delta t$, and $u_h^n = u_h(t_n)$. Assume throughout this paper that a and c are bounded above and below by positive constants, that they are Lipschitz with respect to u, and that the second partial of a with respect to u is bounded. It is known [4, 8] that, if u_h is extended by linear interpolation to all of J,

(1.4)
$$\begin{aligned} \|u - u_h\|_{L^\infty(J; L^2(\Omega))} + h\|u - u_h\|_{L^\infty(J; H^1(\Omega))} &\leq C(u)[\|u\|_{L^\infty(J; H^r(\Omega))} \\ &+ \left\|\frac{\partial u}{\partial t}\right\|_{L^2(J; H^{r-1}(\Omega))} + \left\|\frac{\partial^2 u}{\partial t^2}\right\|_{L^2(J; L^2(\Omega))}](h^r + \Delta t). \end{aligned}$$

Consider the algebraic problems generated by (1.3.b) at each time step. Let $\mathcal{M}_h = \text{span}[w_1, \ldots, w_m]$, $m = m(h)$, and let A^n and C^n be the matrices given by

$$A^n = ((a(u_h^{n-1})\nabla w_j, \nabla w_i)),$$

(1.5)

$$C^n = ((c(u_h^{n-1})w_j, w_i)) .$$

Also, let

(1.6) $$u_h^n = \sum_{j=1}^{m} \mu_j^n w_j$$

and

(1.7) $$\varphi_i^n = (f(u_h^{n-1}), w_i) + \langle g^n, w_i \rangle .$$

Then the matrix form of (1.3.b) is

(1.8) $$(C^n + \Delta t A^n)(\mu^n - \mu^{n-1}) = -\Delta t A^n \mu^{n-1} + \Delta t \varphi^n .$$

Since the coefficients $a(x, u)$ and $c(x, u)$ were evaluated for u_h at the old time level, t_{n-1}, equations (1.8) are linear. Thus, no nonlinear algebraic problem arises; however, the matrix $C^n + \Delta t A^n$ can be expected to change at each step, and, if a direct method is anticipated to be used to solve the equations, a factorization of $C^n + \Delta t A^n$ into an LU-product would be required on every step. In any of the commonly employed elimination methods [5, 6] the factorization dominates quite severely the computation required to produce the solution of $LU\gamma = \delta$, given L and U; thus, it would be very advantageous to find a modification of (1.3.b), or (1.8), such that the matrix corresponding to $C^n + \Delta t A^n$ would be independent of the step number.

Let $c_0 = c_0(x)$ and $a_0 = a_0(x)$ be fixed, subject to conditions to be determined later. Let

(1.9) $$\tilde{c}^n = c_0 - c(u_h^{n-1}), \quad \tilde{a}^n = a_0 - a(u_h^{n-1}).$$

Now, (1.3.b) is equivalent to

(1.10) $$(c_0 \partial u_h^n, \chi) + (a_0 \nabla u_h^n, \nabla \chi) = (f(u_h^{n-1}), \chi) + \langle g^n, \chi \rangle + (\tilde{c}^n \partial u_h^n, \chi) + (\tilde{a}^n \nabla u_h^n, \nabla \chi), \quad \chi \in \mathcal{M}_h ;$$

since the right-hand side contains u_h^n, these equations do not represent an adequate system practically. We expect only first-order convergence in Δt for any essentially backward differencing for the derivative with respect to time; thus, we can try perturbing (1.10) by terms that are formally of magnitude $O(\Delta t)$. We shall shift ∂u_h^n to ∂u_h^{n-1} and ∇u_h^n to ∇u_h^{n-1} on the right-hand side to obtain the evolution equations

$$
\begin{aligned}
(c_0 \partial u_h^n, \chi) + (a_0 \nabla u_h^n, \nabla \chi) &= (f(u_h^{n-1}), \chi) + \langle g^n, \chi \rangle \\
&\quad + (\tilde{c}^n \partial u_h^{n-1}, \chi) + (\tilde{a}^n \nabla u_h^{n-1}, \nabla \chi), \quad \chi \in \mathcal{M}_h .
\end{aligned} \tag{1.11}
$$

Equation (1.11) is a three-level scheme; thus, it is now necessary to specify both u_h^0 and u_h^1 to initiate the recursion. We shall discuss this issue later. The matricial form of (1.11) is, with C^0 and A^0 corresponding to c_0 and a_0, respectively,

$$
(C^0 + \Delta t A^0)(\mu^n - \mu^{n-1}) = -\Delta t A^n \mu^{n-1} + \Delta t \varphi^n + (C^0 - C^n)(\mu^{n-1} - \mu^{n-2}). \tag{1.12}
$$

An outline of the paper is as follows. The first objective will be to find sufficient conditions on the functions c_0 and a_0 such that (1.11) will be stable. It will be shown that the conditions

$$
\begin{aligned}
&\text{(a)} \quad c_0(x) \ge \frac{1}{2}\left(\max_z c(x,z) + \min_w c(x,w)\right), \\
&\text{(b)} \quad a_0(x) \ge \frac{1}{2} \max_z a(x,z),
\end{aligned} \tag{1.13}
$$

are sufficient to insure stability for (1.11) and, in combination with reasonable other hypotheses, convergence of the solution of (1.12) to that of (1.1). Start-up procedures to obtain u_h^0 and u_h^1 will be introduced. The solution of the algebraic equations will be discussed. Finally, we remark that certain extensions and modifications of the method will be published elsewhere; in particular, a second-order correct analogue of (1.11) will be given.

2. Stability. Let

(2.1) $\quad M(x) = \sup_z c(x,z) \leq M_0 < \infty, \quad m(x) = \inf_z c(x,z) \geq m_0 > 0, \quad x \in \Omega.$

No loss of generality results in assuming that $M(x) \leq M_0$ and $m(x) \geq m_0$, provided that $c(x, u(x,t)) \in [2m_0, \frac{1}{2}M_0]$, $0 < m_0 \leq M_0 < \infty$, for all $x \in \Omega$ and $t \in J$ and that, under the assumptions (2.1), u_h converges uniformly to u.

Lemma 2.1. Let $c_0(x) \geq \frac{1}{2}(M(x) + m(x))$, and let $\zeta: \{t_0, \dots, t_N\} \to L^2(\Omega)$. For $2 \leq n \leq N$,

(2.2) $$\sum_{j=2}^{n} [(c_0 \partial\zeta^j, \partial\zeta^j) - (\tilde{c}^j \partial\zeta^{j-1}, \partial\zeta^j)]\Delta t \geq \sum_{j=2}^{n} (m\partial\zeta^j, \partial\zeta^i)\Delta t - (Q\partial\zeta^1, \partial\zeta^1)\Delta t,$$

where $Q(x) = \max(\frac{1}{2}(c_0(x) - m(x)), \frac{1}{4}(M(x) - m(x)))$.

Proof. Note that the left-hand side of (2.2) dominates

$$\sum_{j=2}^{n-1} (\{c_0 - \tfrac{1}{2}|\tilde{c}^j| - \tfrac{1}{2}|\tilde{c}^{j+1}|\}\partial\zeta^j, \partial\zeta^j)\Delta t + (\{c_0 - \tfrac{1}{2}|\tilde{c}^n|\}\partial\zeta^n, \partial\zeta^n)\Delta t - (\tfrac{1}{2}|\tilde{c}^2|\partial\zeta^1, \partial\zeta^1)\Delta t.$$

Now, it is easy to see that $c_0 - |\tilde{c}^j| \geq m$ and that $\frac{1}{2}|\tilde{c}^2| \leq Q$.

Lemma 2.2. If $a_0(x) \geq \frac{1}{2}\sup_z a(x,z)$ and $\zeta: \{t_0, \dots, t_N\} \to H^1(\Omega)$, then, for $2 \leq n \leq N$,

(2.3) $$2\Delta t[(a_0 \nabla\zeta^n, \partial\nabla\zeta^n) - (\tilde{a}^n \nabla\zeta^{n-1}, \partial\nabla\zeta^n)] \geq (a(u_h^{n-1})\nabla\zeta^n, \nabla\zeta^n) - (a(u_h^{n-1})\nabla\zeta^{n-1}, \nabla\zeta^{n-1}).$$

Proof. It is easy to see that

$$\Delta t[(a_0 \nabla\zeta^n, \partial\nabla\zeta^n) - (\tilde{a}^n \nabla\zeta^{n-1}, \partial\nabla\zeta^n)] \geq (\{a_0 - \tfrac{1}{2}|a_0 + \tilde{a}^n|\}\nabla\zeta^n, \nabla\zeta^n) + (\{\tilde{a}^n - \tfrac{1}{2}|a_0 + \tilde{a}^n|\}\nabla\zeta^{n-1}, \nabla\zeta^{n-1}).$$

The condition (1.13.b) implies that $|a_0 + \tilde{a}^n| = 2a_0 - a(u_h^{n-1})$, and (2.3) results.

Lemmas 2.1 and 2.2, when applied to $\zeta^n = u_h^n$, imply a local stability in time for (1.11); global stability will follow from the convergence analysis of the next section.

3. Convergence. The analysis of the convergence of u_h to the solution of (1.1) will utilize techniques due to Dendy [1], Douglas [2], Rachford [7], and Wheeler [8]. Let an elliptic projection of u into $\mathcal{M}_h$ be defined by the map $w_h : \overline{J} \to \mathcal{M}_h$ given by

$$(3.1) \qquad (a(u(t))\nabla(u(t) - w_h(t)), \nabla\chi) + (u(t) - w_h(t), \chi) = 0, \quad \chi \in \mathcal{M}_h .$$

Let $\eta = u - w_h$. It can be shown by using (3.1) and its derivatives with respect to time that, for $a(x,u)$ sufficiently smooth, $0 \le k \le 2$, and $j = 0$ or 1,

$$(3.2) \qquad \left\|\frac{\partial^k \eta}{\partial t^k}\right\|_j \le C(u) \sum_{m=0}^{k} \left\|\frac{\partial^m u}{\partial t^m}\right\|_q \cdot h^{q-j}, \quad 1 \le q \le r,$$

where $C(u)$ depends on certain lower norms of u. For example, if $k = 0$ and $j = 0$, $C(u) = C\cdot(1 + \|u\|_1)$, and if $k = 2$ and $j = 0$, one form of $C(u)$ is given by

$$C(u) = C(\|u\|_{W_3^1(\Omega)}, \left\|\frac{\partial u}{\partial t}\right\|_{L^\infty(\Omega)}, \left\|\frac{\partial u}{\partial t}\right\|_{W_3^1(\Omega)}, \left\|\frac{\partial^2 u}{\partial t^2}\right\|_{L^\infty(\Omega)}, \left\|\frac{\partial^2 u}{\partial t^2}\right\|_{W_3^1(\Omega)}).$$

Let $\xi = w_h - u_h$. A routine but tedious calculation shows that

$$(3.3) \qquad (c_0 \partial\xi^n, \chi) + (a_0 \nabla\xi^n, \nabla\chi) - (\tilde{c}^n \partial\xi^{n-1}, \chi) - (\tilde{a}^n \nabla\xi^{n-1}, \nabla\chi) = (E^n, \chi) + (F^n, \nabla\chi)$$

for $n \ge 2$ and $\chi \in \mathcal{M}_h$, where

$$(3.4) \qquad \begin{aligned} &\text{(a)} \quad E^n = -\eta^n - c(u^n)\partial\eta^n + c(u^n)(\partial u^n - \frac{\partial u^n}{\partial t}) - (c(u^n) - c(u_h^{n-1}))\partial w_h^n \\ &\qquad\qquad + f(u^n) - f(u_h^{n-1}) + \tilde{c}^n(\partial w_h^n - \partial w_h^{n-1}), \\ &\text{(b)} \quad F^n = (a(u_h^{n-1}) - a(u^n))\nabla w_h^n + \tilde{a}^n(\nabla w_h^n - \nabla w_h^{n-1}). \end{aligned}$$

Choose the test function $\chi = \partial\xi^n$ in (3.3). It follows from Lemmas 2.1 and 2.2 that, with $Q(x) \le Q_0$,

$$(3.5) \qquad \begin{aligned} &m_0 \sum_{j=2}^{n} \|\partial\xi^j\|^2 \Delta t - Q_0\|\partial\xi^1\|^2 \Delta t + \frac{1}{2}\sum_{j=2}^{n} [(a(u_h^{j-1})\nabla\xi^j, \nabla\xi^j) - (a(u_h^{j-1})\nabla\xi^{j-1}, \nabla\xi^{j-1})] \\ &\qquad \le \sum_{j=2}^{n} [(E^j, \partial\xi^j) + (F^j, \nabla\partial\xi^j)]\Delta t . \end{aligned}$$

Let

$$(3.6) \qquad \|\chi\|^2_{1,j} = (a(u_h^{j-1})\nabla\chi, \nabla\chi),$$

and note that

$$\|\chi\|^2_{1,j} = \|\chi\|^2_{1,j-1} + ((a(u_h^{j-1}) - a(u_h^{j-2}))\nabla\chi, \nabla\chi).$$

Since we assumed that $a(x, u)$ and $\frac{\partial a}{\partial u}(x, u)$ are Lipschitz continuous with respect to u, $a(u_h^{j-1}) - a(u_h^{j-2}) = \Delta t a_u\{-\partial\xi^{j-1} + \partial w_h^{j-1}\}$. It is then easy to see that

$$(3.7) \qquad \|\chi\|^2_{1,j} = [1 + \Delta t\,\theta_j\{\|\partial\xi^{j-1}\|_{L^\infty(\Omega)} + \|\partial w_h^{j-1}\|_{L^\infty(\Omega)}\}]\|\chi\|^2_{1,j-1},$$

where $|\theta_j|$ is bounded by the Lipschitz constant A for $a(x, \cdot)$. Hence,

$$m_0 \sum_{j=2}^{n} \|\partial\xi^j\|^2 \Delta t + \frac{1}{2}\|\xi^n\|^2_{1,n} \leq Q_0\|\partial\xi^1\|^2 \Delta t + \frac{1}{2}\|\xi^1\|^2_{1,1}$$

$$(3.8) \qquad + A\sum_{j=1}^{n-1}(\|\partial\xi^j\|_{L^\infty} + \|\partial w_h^j\|_{L^\infty})\|\xi^j\|^2_{1,j}\Delta t + \sum_{j=2}^{n}[(E^j, \partial\xi^j) + (F^j, \nabla\partial\xi^j)]\Delta t.$$

In order to estimate the last two terms on the right-hand side, we shall assume certain norms of w_h are bounded. In particular, assume that

$$(3.9) \qquad \|w_h\|_{L^\infty(J; W^1_\infty(\Omega))} + \left\|\frac{\partial w_h}{\partial t}\right\|_{L^\infty(J; W^1_\infty(\Omega))} \leq C_1(u).$$

An inequality of the form (3.9) can be derived from (3.2), smoothness assumptions on u, and inverse hypotheses of the spaces $\mathcal{M}_h$; however, (3.9) should follow from less restrictive conditions when the L^∞-analysis of Galerkin methods for elliptic equations becomes more advanced.

It can be seen from (3.2) that

$$
\begin{aligned}
&\|\partial(w_h^j - w_h^{j-1})\|_q \le C(u)\|u\|_{H^2(J;H^1(\Omega))}\Delta t, \qquad q = 0 \text{ or } 1,\\
&\|\partial \nabla w_h^j\|\Delta t \le C(u)\|u\|_{H^1(t_{j-1},t_j;H^1(\Omega))}(\Delta t)^{1/2},\\
(3.10)\qquad &\|u^j - u^{j-1}\| \le \|\frac{\partial u}{\partial t}\|_{L^2(t_{j-1},t_j;L^2(\Omega))}(\Delta t)^{1/2},\\
&\|\partial u^j - \frac{\partial u^j}{\partial t}\| \le \|\frac{\partial^2 u}{\partial t^2}\|_{L^2(t_{j-1},t_j;L^2(\Omega))}(\Delta t)^{1/2}.
\end{aligned}
$$

Thus, (3.4.a) and (3.10) imply that

$$
\begin{aligned}
&\|(E^j, \partial\xi^j)| \le \epsilon\|\partial\xi^j\|^2 + C\|E^j\|^2\\
(3.11)\qquad &\le \epsilon\|\partial\xi^j\|^2 + C(u)\{\|\eta^j\|^2 + \|\eta^{j-1}\|^2 + \|\partial\eta^j\|^2\\
&\quad + \|\xi^{j-1}\|^2 + \|u\|^2_{H^2(J;H^1(\Omega))}(\Delta t)^2\}.
\end{aligned}
$$

Also, it follows from (3.4.b) and (3.10) that

$$
(3.12)\qquad \|F^j\| \le C(u)\{\|\xi^{j-1}\| + \|\eta^{j-1}\| + \|u\|_{H^2(J;H^1(\Omega))}\Delta t\}.
$$

Another calculation shows that

$$
\begin{aligned}
&\|\partial F^j\| \le C(u)\{\|\partial\eta^{j-1}\| + \|\eta^{j-1}\| + \|\partial\xi^{j-1}\| + \|\xi^{j-1}\|\\
(3.13)\qquad &\quad + \|u\|_{H^2(J;H^1(\Omega))}\Delta t\}.
\end{aligned}
$$

Now, recall that

$$
\sum_{j=2}^{n}(F^j, \nabla\partial\xi^j)\Delta t = (F^n, \nabla\xi^n) - (F^2, \nabla\xi^1) - \sum_{j=3}^{n}(\partial F^j, \nabla\xi^{j-1})\Delta t .
$$

Assume that the solution u of (1.1) satisfies the following restrictions:

$$
(3.14)\qquad u \in H^1(J;H^r(\Omega)) \cap W^2_\infty(J;W^1_3(\Omega)) .
$$

Then, there exists a constant $K(u)$, depending on the norms of u implied in (3.14) (which cover those appearing in the various $C(u)$'s of (3.2)) such that

$$\sum_{j=2}^{n} \|\partial\xi^j\|^2\Delta t + \|\xi^n\|_1^2$$

(3.15) $$\leq C(u)\{\sum_{j=1}^{n-1}(1 + \|\partial\xi^j\|_{L^\infty(\Omega)})\|\xi^j\|_1^2\Delta t + \|\xi^n\|^2 + \|\xi^{n-1}\|^2 + \|\xi^1\|_1^2 + \|\partial\xi^1\|^2\Delta t\}$$
$$+ K(u)\{h^{2r} + (\Delta t)^2\} ;$$

(3.15) results from (3.2), (3.8)-(3.14), and the equivalence of $\|\cdot\|_1$ and $\|\cdot\|_{1,j} + \|\cdot\|$. The terms $\|\xi^n\|^2$ and $\|\xi^{n-1}\|^2$ must be covered. Since

(3.16) $$\|\xi^k\|^2 \leq \|\xi^1\|^2 + \epsilon\sum_{j=2}^{k}\|\partial\xi^{j-1}\|^2\Delta t + C\sum_{j=1}^{k}\|\xi^j\|^2\Delta t ,$$

if the assumption is made that the startup procedure produces u_h^0 and u_h^1 such that

(3.17) $$\|\xi^1\|_1^2 + \|\partial\xi^1\|^2\Delta t \leq K(u)\{h^{2r} + (\Delta t)^2\},$$

then, with adjusted $C(u)$ and $K(u)$,

(3.18) $$\sum_{j=2}^{n}\|\partial\xi^j\|^2\Delta t + \|\xi^n\|_1^2 \leq C(u)\sum_{j=1}^{n-1}(1 + \|\partial\xi^j\|_{L^\infty(\Omega)})\|\xi^j\|_1^2\Delta t + K(u)\{h^{2r} + (\Delta t)^2\}.$$

The following lemma is trivial.

Lemma 3.1. Let $f_j \geq 0$, $\beta_j \geq 0$, and $\gamma > 0$. Assume that, for $n = 1,\ldots,m$,

$$f_n \leq \sum_{j=1}^{n-1}\beta_j f_j \Delta t + \gamma$$

and

$$\sum_{j=1}^{n-1}\beta_j\Delta t \leq M$$

Then, $f_n \leq \gamma\exp(M)$, $n = 1,\ldots,m$.

In order to bound ξ from (3.18), make the induction hypothesis that

(3.19) $$\sum_{j=1}^{n-1}\|\partial\xi^j\|_{L^\infty(\Omega)}\Delta t < 1;$$

then it follows from (3.18), (3.19), and Lemma 3.1 that

(3.20) $$\sum_{j=1}^{n}\|\partial\xi^j\|^2\Delta t + \|\xi^n\|_1^2 \leq \exp\{(T+1)C(u)\}K(u)[h^{2r} + (\Delta t)^2].$$

Then, it follows from (3.20) and the inverse hypothesis (1.2.b) that

$$(3.21)\qquad \Big(\sum_{j=1}^{n} \|\partial\xi^j\|_{L^\infty(\Omega)} \Delta t\Big)^2 \le K_1(u)[h^{2r-d} + h^{-d}(\Delta t)^2],$$

and the right-hand side will tend to zero as $h \to 0$ if

$$(3.22)\qquad r > \frac{1}{2}d \quad\text{and}\quad \Delta t = ch^r ,$$

which justifies the induction hypothesis. Thus, the following theorem has been proved.

Theorem 3.2. Let the solution u of (1.1) satisfy (3.14) for some $r > \frac{1}{2}d$, and let $\mathcal{M}_h$ satisfy (1.2) for the same r. If u_h satisfies (3.17) and is determined by (1.11) for $n \geq 2$ with $a_0(x)$ and $c_0(x)$ satisfying (1.13) and if $\Delta t \leq ch^r$, then there exists a constant $K(u)$, depending on the norms of u required finite by (3.14), such that (with $\zeta^j = u^j - u_h^j$)

$$(3.23)\qquad \sum_{j=2}^{n} \|\partial\zeta^j\|^2 \Delta t + h^2\|\zeta^n\|_1^2 + \|\zeta^n\|^2 \le K(u)[h^{2r} + (\Delta t)^2]$$

for $2 \leq n \leq T(\Delta t)^{-1}$.

Note that piecewise-linear functions $(r = 2)$ over a quasi-regular triangulation can be used for $\mathcal{M}_h$ for Ω of dimension not greater than three. Also, if only optimal order estimates in $H^1(\Omega)$ are wanted, the constraints of (3.14) can be eased somewhat. By modifying the argument leading to Lemma 2.1 slightly, the restriction on $c_0(x)$ can be changed to $c_0(x) \geq \frac{1}{2} M(x) + \delta$ for some $\delta > 0$.

4. Startup procedures. It is necessary to evaluate u_h^0 and u_h^1 in such a way that (3.17) is satisfied. The function u_h^0 can be obtained quite near to w_h^0 by using some reasonable iterative method, such as the conjugate gradient procedure, to approximate the solution of (3.1) at time zero. Then, (1.11) can be adapted as an iterative method to find u_h^1; only two or three iterations would be necessary for (3.17) to hold.

5. Linear equations and operation counts. Consider first the case of two space variables ($d = 2$). George [5] has shown in some special cases that, if $m = m(h)$ is the dimension of $\mathcal{M}_h$, the factorization of $C^0 + \Delta t A^0$ requires $O(m^{3/2})$ operations and that the solution of (1.12), given the factorization, requires $O(m \log m)$ operations. Hoffman, Martin, and Rose [6] have shown that such bounds are minimal. If we conjecture the validity of the bounds, then with $\Delta t = h^r = O(m^{-r/2})$ the total number of arithmetic operations needed is $O(m^{(r+2)/2} \log m)$, which is nearly best possible for a first-order correct-in-time method (since the solution is defined by $O(m^{(r+2)/2})$ parameters).

For $d = 3$, George's nested dissection method seems to require $O(m^2)$ operations for the factorization and $O(m^{4/3})$ for each time step. Now, $\Delta t = O(m^{-r/3})$ and the total work becomes $O(m^{(r+4)/3})$ operations, against an optimal estimate of $O(m^{(r+3)/3})$ for a first-order method. Apparently, there are no known rigorous estimates for the factorization and a single solution of equations of the form (1.12).

References

1. J. Dendy, An analysis of some Galerkin schemes for the solution of nonlinear time-dependent problems, SIAM J. Numer. Anal. 12(1975), 541-565.

2. J. Douglas, Jr., Survey of numerical methods for parabolic differential equations, Advances in Computers, vol. II, Academic Press, New York, 1961.

3. J. Douglas, Jr., and T. Dupont, Galerkin methods for parabolic equations, SIAM J. Numer. Anal. 7(1970), 575-626.

4. T. Dupont, L_2 error estimates for projection methods for parabolic equations in approximating domains, Mathematical Aspects of Finite Elements in Partial Differential Equations, Academic Press, New York, 1974.

5. A. George, Nested dissection on a regular finite element mesh, SIAM J. Numer. Anal. 10(1973), 345-363.

6. A. J. Hoffman, M. S. Martin, and D. J. Rose, Complexity bounds for regular finite difference and finite element grids, SIAM J. Numer. Anal. 10(1973), 364-369.

7. H. H. Rachford, Jr., Two-level discrete-time Galerkin approximations for second order nonlinear parabolic partial differential equations, SIAM J. Numer. Anal. 10(1973), 1010-1026.

8. M. F. Wheeler, A priori L_2 error estimates for Galerkin approximations to parabolic partial differential equations, SIAM J. Numer. Anal. 10(1973), 723-759.

AN AUTOMATIC ONE-WAY DISSECTION ALGORITHM FOR IRREGULAR FINITE ELEMENT PROBLEMS*

Alan George

1. Introduction

In this paper we consider the problem of directly solving the N by N system of linear equations

(1.1) $Ax = b$,

where A is a sparse N by N positive definite matrix arising in the application of finite element methods [15, 17]. The method we use is standard; the matrix A is factored into the product LL^T, where L is lower triangular, and then the triangular systems $Ly=b$ and $L^Tx=y$ are solved to obtain x.

When the matrix A is factored it usually suffers fill; that is, making the usual assumption that exact cancellation does not occur, $L+L^T$ is usually fuller than A. Since PAP^T is positive definite for any N by N permutation matrix P, Cholesky's method can be used to solve the equivalent system

(1.2) $(PAP^T)(Px) = Pb$.

It is well known that a judicious choice of P can often drastically reduce fill and/or arithmetic requirements.

Recently, the author has described two efficient orderings for the system of $N=n^2$ equations arising in connection with the use of finite difference or finite element methods on an n by n grid [6]. These schemes, called nested dissection and one-way dissection, are efficient in the sense that they reduce the arithmetic required to factor A to $O(N^{3/2})$ and $O(N^{7/4})$ respectively, compared to $O(N^2)$ if the usual row by row numbering of the grid is used. In addition, the amount of storage required if one uses these dissection strategies is $O(N \log N)$ and $O(N^{5/4})$, compared to $O(N^{3/2})$. In [6], it was demonstrated that by careful selection of data structures, these orderings could be utilized in linear equation solvers so that their execution times and storage requirements were as the operation and fill counts suggest.

However, although simple model n by n grid problems are easy to analyze and interesting from a theoretical point of view, a much more desirable practical requirement is to find similarly efficient orderings for less regular problems. In [8], George and Liu have provided an automatic algorithm for producing nested dissection orderings for irregular finite element problems. The objective of this paper is to provide an automatic scheme for finding orderings analogous to the one-way dissection orderings. As we shall see later, the storage requirements for these orderings appear to grow as $N^{5/4}$, as we expect in view of the results for the n by n grid problem. On the surface, it appears that such orderings are inferior to nested dissection orderings, whose storage requirements only grow as $O(N \log N)$. However, the

*Research supported in part by Canadian National Research Council grant A8111.

estimates are asymptotic, and unless N is very large indeed, the one-way dissection orderings appear to require considerably less storage than the nested dissection orderings. (This is also the situation for the n by n grid problem.) In exchange for lower storage requirements, we normally perform more arithmetic than for the nested dissection orderings. Thus, the automatic determination of such one-way orderings for irregular problems is important when storage is limited. In this paper a heuristic algorithm is described for finding one-way dissection orderings for sparse matrix problems, and some numerical experiments describing its application to some finite element problems are provided.

The class of problems for which this one-way dissection algorithm is primarily intended are those arising in connection with the application of finite difference and finite element methods. For two dimensional problems, they can be characterized as follows. Let M be a planar mesh consisting of triangles and/or quadrilaterals called elements, as shown in Figure 1.1, where adjacent elements have a common side or vertex. The mesh has nodes at each vertex, and there may also be nodes lying on element edges and faces; associated with each node is one or more variables x_i, and for some labelling of these N variables we define a finite element system $Ax = b$ associated with M as one for which A is symmetric and positive definite, and for which $A_{ij} \neq 0 \Rightarrow x_i$ and x_j are associated with the same node, or nodes belonging to the same element.

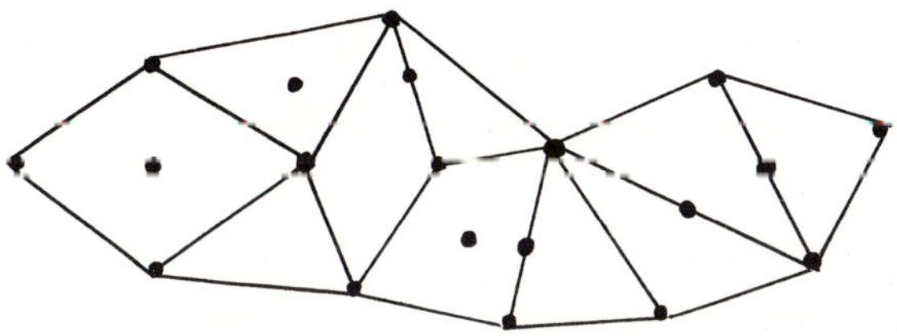

Figure 1.1 A 20 node finite element mesh with 10 elements.

An outline of the paper is as follows. In section 2 a simple model grid problem is analyzed to provide the motivation for the ordering algorithm. Section 3 contains a description of the actual ordering algorithm, and section 4 contains some numerical experiments with the algorithm, applied to problems typical of those arising in finite element applications. Finally, section 5 contains some concluding remarks.

2. Motivation†

Consider an m by ℓ grid or mesh as shown in Figure 2.1, having $N=m\ell$ nodes. The corresponding matrix problem we consider has the property that for any numbering of the nodes from 1 to N, the coefficient matrix A satisfies $A_{ij} \neq 0 \Rightarrow$ node i and node j belong to the same small square. This corresponds to the familiar 9 point difference

†This development closely parallels that in George [6, section 4] except that here we treat the case for $m \neq \ell$. This is important as it provides a mechanism for choosing α when the mesh is irregular.

applied to a regular discretization of a rectangular domain.

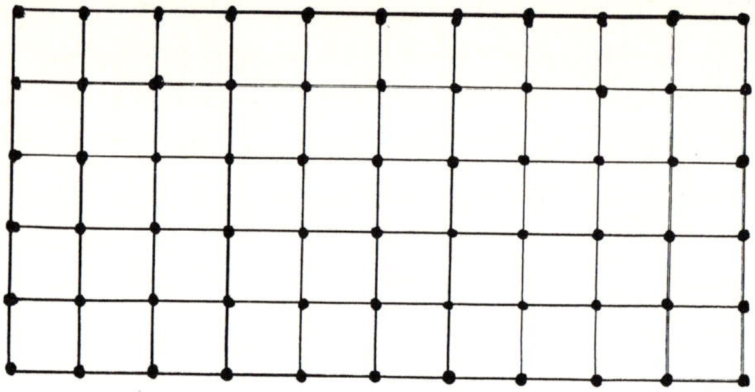

Figure 2.1   An m by ℓ grid with m=6 and ℓ=11.

Let α be an integer satisfying $1 \le \alpha \le \ell$, and choose α vertical grid lines (separators) which dissect the grid into α+1 independent blocks, as depicted in Figure 2.2, where α=3. The α+1 independent blocks are numbered row by row, followed by the α separators, as indicated in Figure 2.2. The matrix structure that this ordering induces is shown in Figure 2.3.

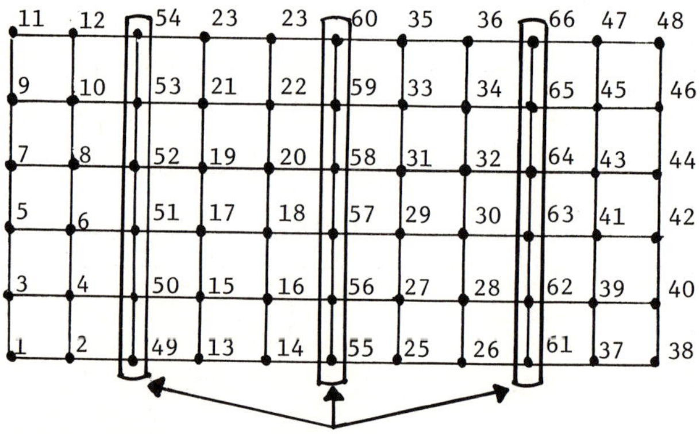

Figure 2.2   One-way dissection ordering of the grid of Figure 2.1, with the number of separators α equal to 3.

The keys to the efficient use of this ordering can be illustrated by considering the block 2 by 2 symmetric positive definite matrix

(2.1)
$$A = \begin{pmatrix} A_{11} & A_{12} \\ A_{12}^T & A_{22} \end{pmatrix},$$

whose (block) Cholesky factorization is

(2.2)
$$L = \begin{pmatrix} L_{11} & 0 \\ W_{12}^T & L_{22} \end{pmatrix},$$

where

(2.3) $$A_{11} = L_{11}L_{11}^T ,$$

(2.4) $$W_{12} = L_{11}^{-1} A_{12} ,$$

and

(2.5) $$L_{22}L_{22}^T = \tilde{A}_{22} = A_{22} - A_{12}^T A_{11}^{-1} A_{12} .$$

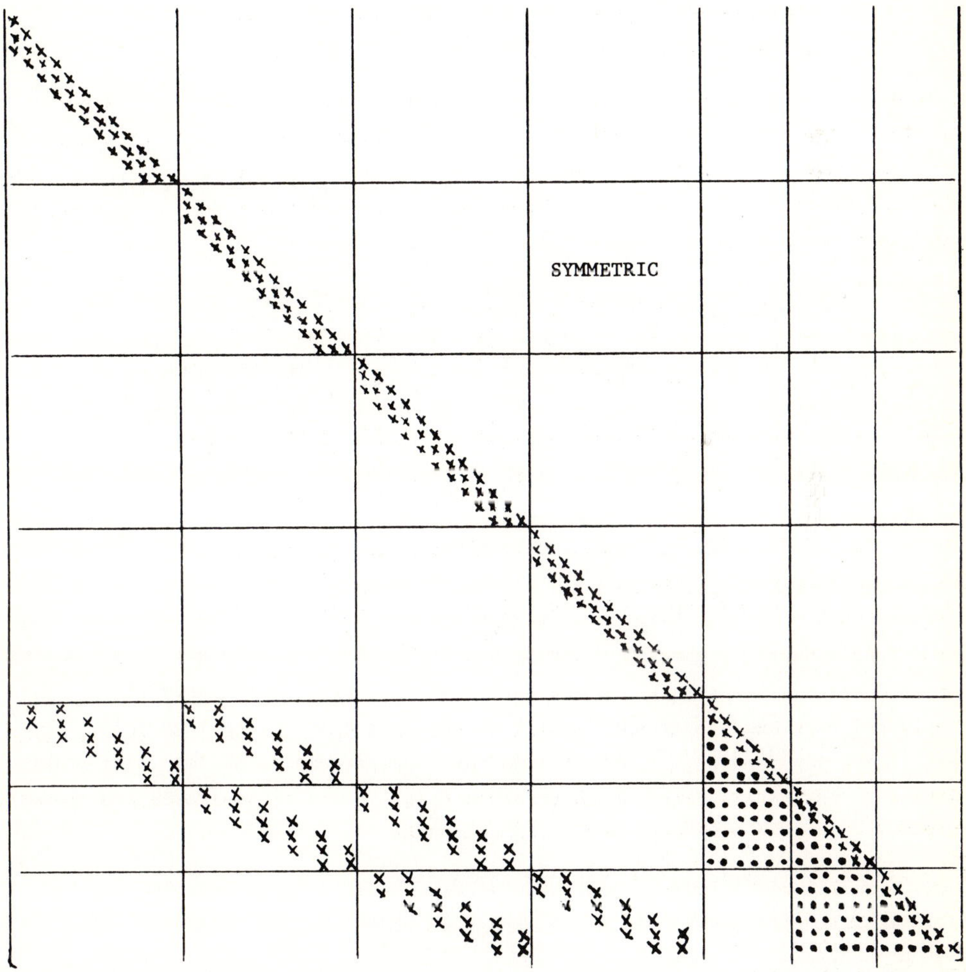

Figure 2.3 Matrix structure induced by the ordering of Figure 2.2. The dots indicate the fill suffered by the diagonal block corresponding to the separators.

As usual, it should be understood that inverses are not actually computed; instead, the appropriate triangular systems are solved.

The first observation is that $\tilde{A}_{22}$ can be computed either as $A_{22}-(A_{12}^T L_{11}^{-T})(L_{11}^{-1} A_{12}) = A_{22} - W_{12}^T W_{12}$, or as $A_{22} - A_{12}^T (L_{11}^{-T} (L_{11}^{-1} A_{12}))$. In general, for sparse matrices these factorizations require different amounts of computation. A particularly important point is that the second form of the computation does not require storage for W_{12}, since the computation can be carried out one column at a time. Hence only one auxiliary vector is required. On the other hand, the first computation appears to require the storage of the entire matrix W_{12}. See George [5] for details.

The second observation is that we may not wish to retain W_{12}. As observed by Bunch and Rose [3], it may require fewer arithmetic operations to operate only implicitly with W_{12}, by using (2.4). For example, to compute $\tilde{x}_2 = W_{12}x_2$ we can calculate $z = A_{12}x_2$, and then solve the triangular system $L_{11}\tilde{x}_2 = z$. If there are fewer nonzeros in total in L_{11} and A_{12} than in W_{12}, this mode of calculating will save arithmetic operations, and will also save primary storage, since A_{12} rather than W_{12} is stored. Here primary storage is that actually used to store numerical values, rather than storage for pointers etc, which we will call overhead storage.

Returning now to our one-way dissection ordering, it turns out to be crucial to use both observations, where the leading $\alpha+1$ blocks corresponding to the "independent blocks" are viewed as A_{11}, and the diagonal block corresponding to the α separators is A_{22}. Thus, the only part of L that we store will be the lower triangles of the $2\alpha+1$ diagonal blocks, along with the $\alpha-1$ m by m "fill" blocks directly below the last α diagonal blocks. The computation of $\tilde{A}_{22}$ is performed in the asymmetric way, thus avoiding storing W_{12} at any time.

We now do a rough analysis of storage requirements in order to determine what value α should have for general m and ℓ. In the actual implementation the diagonal blocks A_{11} and A_{22} are treated as one matrix whose variation in bandwidth is exploited by a storage scheme similar to that proposed by Jennings [11]. The matrix A_{12} is very sparse, so only its nonzeros are stored. The important point here is that the total number of nonzeros in L_{11}, L_{22} and A_{12} is given approximately by

$$(2.6)\quad p(\alpha) = \frac{m(\ell-\alpha)^2}{(\alpha+1)} + \frac{m(m+1)}{2} + (\alpha-1)m^2 + O(\alpha m, \ell).$$

Differentiating with respect to α, we find p is approximately minimized for

$$(2.7)\quad \tilde{\alpha} = \ell\{2/3(m+1)\}^{\frac{1}{2}} - 1,$$

yielding

$$(2.8)\quad p(\tilde{\alpha}) = \sqrt{3/2}\, m\ell\, \{\ell^{\frac{1}{2}}+m^{\frac{1}{2}}\} + O(m\ell).$$

Of course in practice $\tilde{\alpha}$ must be chosen to be an integer, and the dissection in general will not be exactly uniform unless $(\ell-\alpha)\ /\ (\alpha+1)$ is also an integer. However, these minor observations turn out to be irrelevant from a practical viewpoint since $|p'(\alpha)|$ is small near $\tilde{\alpha}$.

Note that if we used an ordinary column-by-column numbering scheme we would obtain a band matrix having bandwidth $\beta = m+2$. (We define the bandwidth β of a matrix M to be $\max \{|i-j| \mid M_{ij} \neq 0\}$.) Using a standard band storage scheme [16], our storage requirements would be $\frac{1}{2}m^2\ell + O(m\ell)$. Comparing with (2.8), we can conclude that our one-way dissection scheme will be preferable to the standard band scheme provided that $m \gtrsim \sqrt{6\ell}$. When m is proportional to ℓ, (2.8) implies that the one-way dissection scheme requires $O(N^{5/4})$ storage, compared to $O(N^{3/2})$ for the standard band scheme. It is straightforward exercise to show that the operation count for the factorization is $O(N^{7/4})$ compared to $O(N^2)$ for the standard band scheme.

This simple analysis provides us with some insight into the general nature of a one-way dissection ordering for irregular mesh problems such as the one depicted in Figure 1.1. We would like to find a set of separators, hopefully having relatively few nodes, which disconnect the mesh into pieces which can be numbered so that the corresponding matrices have a small bandwidth or envelope. In the next section we rephrase the ordering problem in graph theory terms and provide a heuristic algorithm for finding such separators and orderings.

3. The Ordering Algorithm

3.1 Graph Theory Terminology

It is convenient to describe the ordering algorithm in terms of labelling an undirected graph, so we begin by introducing some standard graph theory notations. An undirected graph $G = (X,E)$ is a finite non-empty set X of nodes or vertices together with a set E of edges, which are ordered pairs of distinct nodes of X. A graph $G' = (X',E')$ is a <u>subgraph</u> of G if $X' \subseteq X$ and $E' \subseteq E$. For $Y \subset X$, the <u>section graph</u> $G(Y)$ is the subgraph $(Y,E(Y))$, where

$$E(Y) = \left\{ \{x,y\} \in E \mid x \in Y,\ y \in Y \right\}$$

Two nodes x and y are adjacent if $\{x,y\} \in E$. For $Y \subset X$, the adjacent set of Y, denoted by $\mathrm{Adj}(Y)$, is

$$\mathrm{Adj}(Y) = \left\{ \{x \in X \backslash Y \mid \{x,y\} \in E \text{ for some } y \in Y \right\}.$$

For distinct nodes x and y in G, a path from x to y of length k is an ordered set of distinct nodes $(v_1, v_2, \ldots, v_{k+1})$, where $x=v_1$ and $y=v_{k+1}$, such that $v_{i+1} \in \mathrm{Adj}(v_i)$, $i=1,2,\ldots,k$. A graph G is <u>connected</u> if for every pair of distinct nodes x and y, there is at least one path from x to y. If G is <u>disconnected</u>, it consists of two or more <u>connected components</u>.

The set $Y \subset X$ is a separator of the connected graph G if the section graph $G(X\backslash Y)$ is disconnected; Y is a minimal separator if no proper subset of Y is a separator of G.

A partitioning P of the graph G is a subset of the power set of X:

$$P = \{Y_1, Y_2, \ldots, Y_p\},$$

where $\bigcup_{i=1}^{p} Y_i = X$ and $Y_i \cap Y_j = \emptyset$ for $i \neq j$. An ordering (labelling, numbering) of G is a bijective mapping α: $\{1,2,3,\ldots,N\} \to X$, where $N=|X|$.

An important type of graph partitioning, which plays an integral role in our one-one-way dissection algorithm, is the class of level structures [1]. A level structure of a connected graph $G = (X,E)$ is a partitioning

$$\mathcal{L} = \{L_0, L_1, \ldots, L_\ell\}$$

of the node set X such that $\mathrm{Adj}(L_0) \subset L_1$, $\mathrm{Adj}(L_\ell) \subset L_{\ell-1}$, and $\mathrm{Adj}(L_i) \subset L_{i-1} \cup L_{i+1}$ for $0 < i < \ell$. A rooted level structure, rooted at $x \in X$, is the level structure

$$\mathcal{L}(x) = \{L_0(x),\ L_1(x), \ldots, L_{\ell(x)}(x)\},$$

where $L_0(x) = \{x\}$ and $L_i(x) = \mathrm{Adj}\left(\bigcup_{j=0}^{i-1} L_j\right)$. The number $\ell(x)$ is sometimes called the eccentricity of x. The diameter of G is then defined by

$$\delta(G) = \max\{\ell(x) \mid x \in X\}.$$

A peripheral node x is one such that $\ell(x) = \delta(G)$.

The effectiveness of the ordering algorithm described later in this section hinges on finding a relatively "long, narrow" level structure. That is, one having relatively many levels, and relatively few nodes in each level. It is intuitively clear that a level structure rooted at a peripheral node would be a good candidate.

Unfortunately, no efficent algorithm is known to determine such nodes for a general graph. Recently Gibbs et. al. [10] have devised an algorithm for finding nodes of high eccentricity. A modification of this algorithm, described in [7], is used in the algorithm described in this paper. The root used for the level structure will be called a pseudo-peripheral node.

We now establish a connection between graphs and matrices. Let A be an N by N symmetric matrix. The labelled undirected graph of A, denoted by $G^A = (X^A, E^A)$, is one for which X^A is labelled from 1 to N and $\{x_i, x_j\} \in E^A$ if and only if $A_{ij} \neq 0$, $i > j$. The unlabelled graph of A is simply G^A with its labels removed. For any N by N permutation matrix P, the unlabelled graphs of A and PAP^T are the same, but the associated labellings differ. Finding a good ordering for A can thus be viewed as finding a good labelling for its graph.

3.2 Description of the Algorithm

We begin with a step-by-step description of the algorithm, followed by some explanatory remarks for the more important steps.

1. Find a pseudo-peripheral node x, using the algorithm described in [7], and then generate the level structure $\mathcal{L}(x) = \{L_1, L_2, \ldots, L_\ell\}$ rooted at x.
2. (Estimate $\tilde{\alpha}$) Calculate $m = N/(\ell+1)$, and if $m \le (6\ell)^{\frac{1}{2}}$, go to step 5. Otherwise set $\tilde{\alpha} = \ell(2/3(m+1))^{\frac{1}{2}} - 1$ and go to step 3.
3. Set $\delta = \ell/(\tilde{\alpha}+1)$. If $\delta \ge 2$, go to step 6. Otherwise go to step 4.
4. (Correct estimate for $\tilde{\alpha}$) Decrement $\tilde{\alpha}$ by 1, and if $\tilde{\alpha} > 1$, go to step 3. Otherwise go to step 5.
5. Set k=1, $Y_1 = X$, $\alpha=0$, and then go to step 8.
6. (Find separators) Set $\alpha \quad = \lfloor\tilde{\alpha}\rfloor$.

For $i = 1,2,\ldots,\alpha$ do the following:

6.1 Set $j = \lfloor i\,\delta - 1 \rfloor$.

6.2 Choose $T_i \subseteq L_j$ to be a minimal separator of G.

7. Let Y_i, $i = 1,2,\ldots,k$ be the connected components of the section graph $G(X \setminus \bigcup_{j=1}^{\alpha} T_j)$, and $Y_{k+j} = T_j$, $j = 1,2,\ldots,\alpha$.

8. Number each Y_i, $i = 1,2,\ldots,k$ consecutively using the reverse Cuthill-McKee (RCM) algorithm [12], and if $\alpha > 0$, number the Y_{k+j}, $j = 1,2,\ldots,\alpha$ consecutively in any order.

Step 1 of the algorithm produces the (hopefully) long, narrow level structure referred to in section 2. This is desirable because the separators are selected as subsets of some of the levels L_i.

The calculation of the numbers m and $\tilde{\alpha}$ computed in step 2 is motivated directly by the crude analysis of the m by ℓ grid in section 2. Since m is the average number of nodes per level, it serves as a measure of the width of the level structure.

Steps 4 and 5 are normally not executed; they are designed to handle anomalous situations where $m \ll \ell$, or when N is simply too small to make the application of the dissection algorithm sensible. In these cases, the entire graph is processed as one block (α=0). That is, an ordinary band type ordering is produced for the graph.

Step 6 performs the actual selection of the separators, and is done essentially as though the graph corresponded to an m by ℓ grid as studied in section 2. As noted earlier, each L_i of $\mathcal{L}$ is a separator of G, although not necessarily minimal. In step 6, α approximately equally spaced levels are chosen from $\mathcal{L}$, and subsets of these levels (the T_i) which are minimal separators are then found. These nodes together correspond to the A_{22} of section 2.

Finally, in step 8 the $k \geq \alpha+1$ independent blocks created by removing the separators from the graph are numbered, using the well known reverse Cuthill-McKee algorithm [12].

Although the choice of α and the method of selection of the separators seems remarkably crude, we have found that attempts at more sophistication do not really yield significant benefits (except for some unrealistic, contrived examples). Just as in the regular rectangular grid case, the storage requirement, as a function of α, is very flat near its minimum. Even relatively large perturbations in the value of α, and in the selection of the separators, produced rather small changes in storage requirements.

4. Some Implementation Details and Numerical Experiments

The linear equations solver used essentially regards the matrix as 2 by 2 partitioned, with the independent blocks formed by the removal of the separators forming the first partition, and the equations corresponding to the separators themselves comprising the second partition. Using the notation of (2.1) - (2.5), the linear equations solver stores L_{11} and L_{22} using a scheme similar to that proposed by Jennings [11], which exploits the variation in bandwidth. The matrix W_{12} is of course not stored; instead, the nonzeros of A_{12} are stored column by column in consecutive locations in a single one dimensional array, with a parallel array containing their row subscripts, and a pointer array indicating the position of the beginning of each column.

In order to gain some insight into the asymptotic behaviour of the ordering algorithm and the quality of the ordering produced, the ordering algorithm/solver combination was applied to problems derived from the graded L mesh shown in Figure 4.1, subdivided by increasing subdivision factors s, yielding s^2 as many triangles as in the original mesh. The numerical results are contained in Tables 4.1 and 4.2.

In Table 4.1, the column "storage" includes <u>all</u> array storage used by the program, including storage for pointers, auxiliary storage, space for the right hand side b, etc. Note that the execution time of the ordering algorithm appears to grow linearly with N. The execution time associated with finding where the fill occurs, so that space for L_{11} and L_{22} can be allocated also appears to require time proportional to N. A description of this algorithm would take us too far afield; details can be found in [9]. Finally, note that the storage requirements grow as $N^{5/4}$, as the results of section 2 suggest.

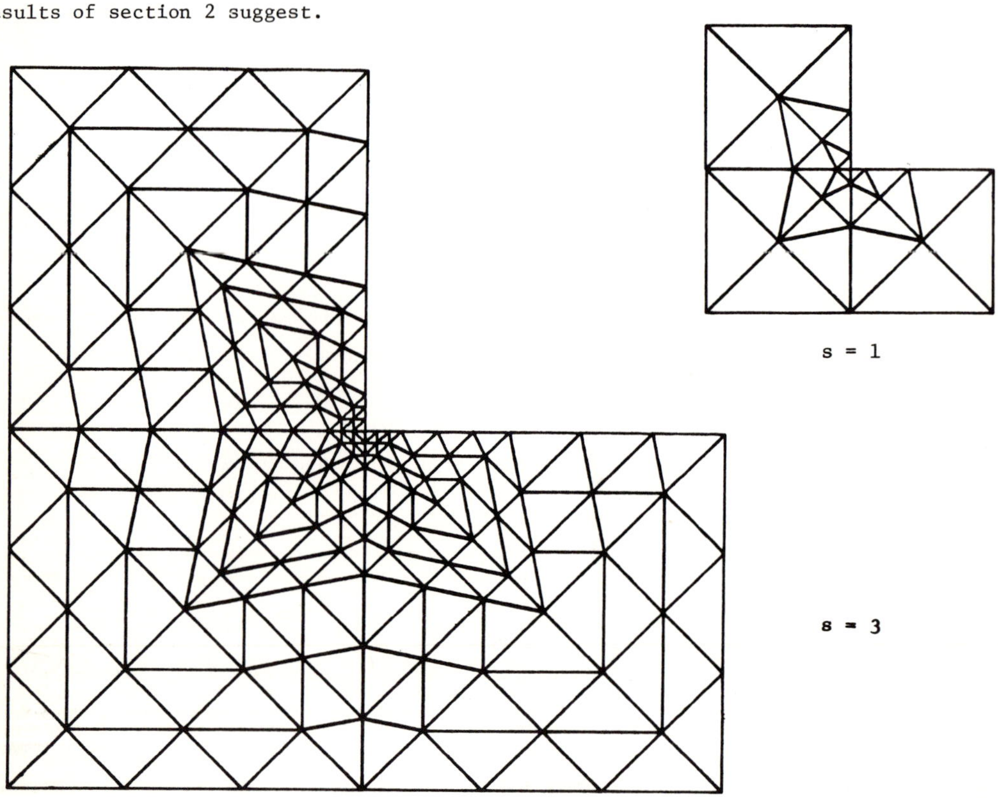

Figure 4.1 Graded L mesh.

In Tables 4.1 and 4.2 we distinguish between (1) execution times due to the determination of the ordering and the allocation of storage, (2) the time required to compute the factorization, and (3) the time required to compute x, given the factorization. In some situations many different problems having the same zero-nonzero structure must be solved, so it makes sense to ignore the cost of finding the ordering and setting up the data structures when evaluating the method. In other cases, many systems having the same coefficient matrix but different right hand sides must be solved. In this situation it may be sensible to evaluate the method on the basis of its solution time alone, and ignore ordering, allocation, and factorization times. All execution times reported are in seconds on an IBM 360/75. The programs are written in Fortran, and were compiled using the optimizing version of the compiler (OPT = 2).

The execution time of the factorization appears to be proportional to $N^{7/4}$, and the solution time appears to grow as $N^{5/4}$. These empirical results are not surprising in view of the results of section 2.

In order to demonstrate that such one-way dissection orderings do have a place in the "sophistication spectrum" of sparse matrix ordering schemes, we include in Tables 4.3 and 4.4 some further results on the graded L problem. The heading RCM refers to the results obtained using the reverse Cuthill-McKee ordering algorithm coupled with a standard band-oriented solver which exploits variation in the bandwidth. We regard this as the standard scheme. The heading NESTED represents the other end of the spectrum, and refers to results obtained by using a slightly revised version of the automatic nested dissection algorithm and solver described in George and Liu [8].

Table 4.1 Performance statistics of the ordering algorithm, storage allocator, and storage requirements, for the graded - L problem and s = 4(1)12.

s	N	Order Time	Order Time / N	Allocation Time	Alloc. Time / N	Storage	Storage / $N^{5/4}$
4	265	.18	6.79(-4)	.07	2.64(-4)	3486	3.26
5	406	.27	6.65(-4)	.11	2.71(-4)	5675	3.11
6	577	.38	6.58(-4)	.16	2.77(-4)	8581	3.03
7	778	.52	6.68(-4)	.21	2.69(-4)	12346	3.00
8	1009	.67	6.64(-4)	.27	2.67(-4)	16667	2.93
9	1270	.84	6.61(-4)	.35	2.76(-4)	21847	2.88
10	1561	1.03	6.60(-4)	.43	2.75(-4)	27860	2.83
11	1882	1.24	6.58(-4)	.52	2.76(-4)	34915	2.81
12	2233	1.47	6.58(-4)	.62	2.77(-4)	42636	2.77

Table 4.2 Performance statistics for the linear equations solver for the graded L problem, with s = 4(1)12.

s	N	Factorization Time	Fact. Time / $N^{7/4}$	Solution Time	Soln. Time / $N^{5/4}$
4	265	.85	4.88(-5)	.09	8.42(-5)
5	406	1.64	4.47(-5)	.14	7.68(-5)
6	577	2.89	4.25(-5)	.21	7.43(-5)
7	778	4.66	4.07(-5)	.29	7.06(-5)
8	1009	7.11	3.94(-5)	.40	7.03(-5)
9	1270	10.07	3.73(-5)	.51	6.73(-5)
10	1561	14.50	3.74(-5)	.65	6.62(-5)
11	1882	20.10	3.72(-5)	.80	6.45(-5)
12	2233	26.69	3.68(-5)	.98	6.38(-5)

Table 4.3 Ordering times for the RCM, one-way dissection and nested dissection algorithm, and the storage requirements for their respective solvers, for the graded - L problem.

s	N	ORDERING TIME			TOTAL STORAGE		
		RCM	ONE-WAY	NESTED	RCM	ONE-WAY	NESTED
4	265	.12	.18	.36	4279	3486	5433
5	406	.19	.27	.61	7764	5675	9135
6	577	.27	.38	.90	12748	8581	13845
7	778	.36	.52	1.27	19497	12346	19863
8	1009	.49	.67	1.72	28277	16667	26691
9	1270	.60	.84	2.25	39354	21847	35252
10	1561	.73	1.03	2.89	52994	27860	44957
11	1882	.88	1.24	3.54	69463	34915	55924
12	2233	1.04	1.47	4.35	89027	42638	68244

because of differences in the coefficients of the estimates and differences in data structure complexity, the one-way scheme requires less storage until N is very large indeed. The one-way dissection ordering algorithm takes about fifty percent more time than the RCM algorithm, but it is considerably faster than the nested dissection algorithm.

Perhaps the most noteworthy aspect of Table 4.4 is that it illustrates how dangerous it is to conclude very much of a practical nature from a study of operation counts alone. While they correctly predict the <u>trends</u> in execution time, differences in data structure complexity can have an enormous effect on the operation per second output of a linear equations solver. Because the data

		FACTORIZATION						SOLUTION					
		OPERATIONS			TIME			OPERATIONS			TIME		
s	N	RCM	ONE-WAY	NESTED	RCM	ONE-WAY	NESTED	RCM	ONE-WAY	NESTED	RCM	ONE-WAY	NESTED
4	265	.297	.473	.330	.34	.90	.69	.749	.651	.738	.06	.08	.11
5	406	.662	1.053	.685	.70	1.71	1.28	1.389	1.159	1.288	.12	.13	.18
6	577	1.288	1.808	1.201	1.27	2.99	2.05	2.318	1.749	1.994	.19	.20	.28
7	778	2.278	3.044	1.988	2.14	4.95	3.09	3.587	2.546	2.923	.29	.28	.38
8	1009	3.749	4.595	3.003	3.38	7.33	4.35	5.251	3.611	4.020	.41	.39	.52
9	1270	5.837	7.159	4.404	5.10	10.33	6.05	7.362	4.726	5.360	.58	.50	.68
10	1561	8.695	10.656	6.113	7.38	14.77	8.07	9.973	6.287	6.893	.77	.65	.87
11	1882	12.490	14.221	8.295	10.40	20.21	10.47	13.139	7.767	8.660	1.01	.80	1.07
12	2233	-	19.620	10.084	-	26.52	13.32	-	9.743	10.631	-	.98	1.30
		$(\times 10^5)$	$(\times 10^5)$	$(\times 10^5)$				$(\times 10^4)$	$(\times 10^4)$	$(\times 10^4)$			

Table 4.4 Operation counts and execution times for the linear equation solvers, for the three different orderings on the graded - L problem.

structures are simplest for the standard scheme, it remains competitive in terms of execution time for values of N much larger than the operation counts would suggest.

There are a number of situations where the one-way dissection scheme could be quite attractive. Since its storage requirements are substantially lower than either of its competitors over a fairly large range of N, in situations where storage is expensive and/or severely restricted, it may be the method of choice even though its execution time may be larger than its competitors. In addition, in some situations involving the solution of some mildly nonlinear or time dependent problems, many systems having the same coefficient matrix must be solved. In these situations, the cost of solving the problem, given the factorization, may be the primary factor governing the method's merit. The last six columns in Table 4.3 indicate that the one-way dissection scheme is a strong contender in these cases.

The one-way dissection approach has the potential to be very important in a multi-processor mini-computer environment. The computation involving the leading α independent blocks can quite conveniently be performed on α different computers, in parallel.

5. Concluding Remarks

We have presented a heuristic algorithm for finding a so-called one-way dissection ordering for an undirected graph. Although the experiments presented are for a single problem, quite extensive testing on a variety of other problems of the same type suggest that for these problems:

1. The ordering algorithm executes in $O(N)$ time.
2. The ordering produced, when used with the solver which exploits the structure of the reordered matrix as described in section 2, yields $(N^{7/4})$ factorization times and $O(N^{5/4})$ storage requirements.

For fairly obvious reasons, the cross-over points where one scheme became more attractive than another (according to some specified criterion) depended upon the particular problem. For example, extremely long slender meshes yielded problems where the ordinary band scheme was uniformly superior for all s. However, for meshes which were undeniably two or three dimensional, there was a substantial range of N where the one-way scheme was, in some respects, the most efficient.

6. References

[1] Arany, I., Smyth, W.F., and Szoda, L. "An improved method for reducing the bandwidth of sparse symmetric matrices", in Information Processing 71: Proceedings of IFIP Congress, North-Holland, Amsterdam, (1972).

[2] Berge, C. The Theory of Graphs and its Applications, John Wiley & Sons Inc., New York, (1962).

[3] Bunch, J.R., and Rose, D.J. "Partitioning, tearing, and modification of sparse linear systems", J. Math. Anal. and Appl., 48, 574-593 (1974).

[4] Duff, I.S. "Sparse matrices", AERE Rept HL 76/485, Harwell, England, Feb 1976.

[5] George, A. "On block elimination for sparse linear systems", SIAM J. Numer. Anal., 11, 585-603 (1974).

[6] George, A. "Numerical experiments using dissection methods to solve n by n grid problems", SIAM. J. Numer. Anal., 14, 161-179 (1977).

[7] George, A., and Liu, J.W.H. "An implementation of a pseudo-peripheral node finder", Dept. of Computer Science Technical Report CS-76-44, University of Waterloo, Oct 1976.

[8] George, A., and Liu, J.W.H. "An algorithm for automatic nested dissection and its application to general finite element problems", Proc. Sixth Conference on Numerical Mathematics and Computing, Winnipeg, Manitoba, 59-94 (1976).

[9] George, A., and Liu, J.W.H. "On finding diagonal block envelopes of triangular factors of partitioned matrices", Dept. of Computer Science Technical Report CS-77-10, University of Waterloo, April 1977.

[10] Gibbs, N.E., Poole, W.G., and Stockmeyer, P.K. "An algorithm for reducing the bandwidth and profile of a sparse matrix", SIAM J. Numer. Anal., 13, 236-250 (1976).

[11] Jennings, A. "A compact storage scheme for the solution of simultaneous equations", Comput. J., 9, 281-285 (1966).

[12] Liu, J.W.H., and Sherman A.H. "Comparative analysis of the Cuthill-McKee and the reverse Cuthill-McKee ordering algorithms for sparse matrices", SIAM J. Numer. Anal., 13, 198-213 (1976).

[13] Parter, S.V. "The use of linear graphs in Gauss elimination", SIAM Rev., 3 364-369 (1961).

[14] Rose, D.J. "A graph theoretic study of the numerical solution of sparse positive definite systems", in Graph Theory and Computing, R.C. Read, editor, Academic Press, (1972).

[15] Strang, G.W., and Fix, G.J. An Analysis of the Finite Element Method, Prentice-Hall Inc., Englewood Cliffs, N.J., (1973).

[16] Wilkinson, J.H., and Reinsch, C. Handbook for Automatic Computation II: Linear Algebra, Springer Verlag, 1971.

[17] Zienkiewicz, O.C. The Finite Element Method in Engineering Science, McGraw-Hill, London, (1970).

GENERALISED GALERKIN METHODS FOR SECOND ORDER EQUATIONS WITH SIGNIFICANT FIRST DERIVATIVE TERMS

A R Mitchell and D F Griffiths

1. Historical

We begin with a brief history of generalised Galerkin methods referred to the problem

$$Au = f \quad \text{in} \quad R$$
$$u = g \quad \text{on} \quad \partial R$$

where A is a second order elliptic operator, not necessarily self-adjoint, and u is the unique solution belonging to an appropriate Hilbert space H. The approximate solution U is given by

$$U = \sum_{i=0}^{N+1} a_i \phi_i$$

where

$$(AU-f, \psi_j) = 0 , \qquad j = 0,1,\ldots,N+1 .$$

We write $\phi_i \in \phi_h$ and $\psi_j \in \psi_h$, where ϕ_h and ψ_h are finite dimensional approximations of H, and are referred to as spaces of trial and test functions respectively.

The method (1.1), with $\psi_i = \phi_i$, was first proposed by the Russian engineer Galerkin. Imprisoned by the Tsar in 1906 he published his first paper on buckling of bars. This method was also used by Bubnov in 1913. Specific examples of the generalised Galerkin method $(\psi_i \neq \phi_i)$, were introduced by Picone in 1928 and Krawchuk in 1932. These were respectively $\psi_i = A\phi_i$ (least squares) and $\psi_i = A'\phi_i$, with A' a differential operator of the same order as A (H^1 Galerkin). In 1934, Slater proposed $\psi_i = \delta_i$ (Dirac delta function) which lead to collocation, and Murray in 1943 proposed a technique similar to the H^{-1} Galerkin method, details of which will be given later. All of these generalised Galerkin methods were unified by Crandall [1] under the title of method of weighted residuals and by Collatz [2] under the heading of error distribution principles. A general survey entitled the Petrov-Galerkin method can be found in Mitchell and Anderssen [3] .

Originally trial and test functions were sought which were global functions and sympathetic to the particular problem being solved. As such they were difficult to define for all but very simple problems (e.g. linear self adjoint operators and square or circular regions). The arrival of the finite element method made the construction of trial and test functions relatively straightforward. Usually they are piecewise polynomials of Lagrange, Hermite, or spline type (Mitchell and Wait [5]). Other trial (test) spaces used are piecewise rationals (Wachspress [4]) for curved boundaries, and piecewise exponentials (Hemker [6] and Griffiths and Lorenz [7]) for boundary layer problems.

2. Statement of Problem

The numerical solution of the convective-diffusion equation

$$\nabla . K \nabla \phi - V . \nabla \phi = 0 \qquad V = (u,v)^T \tag{2.1}$$

presents serious difficulties when the convective term is important [8]. These difficulties arise from the respective elliptic and hyperbolic characters of the two terms and manifest themselves numerically in an oscillatory solution whenever the mesh size exceeds a certain critical value. In (2.1), K is the diffusion coefficient and u and v are known velocity components. Similar difficulties arise with the convective terms in the Navier-Stokes equation in fluid mechanics and with other transport equations.

We consider the simple model problem

$$L\phi \equiv \phi'' - k\phi' = 0 , \qquad \phi(0) = 1 \qquad \phi(1) = 0 \qquad k > 0 . \tag{2.2a}$$

If we define

$$\phi = u(x) + 1 - x ,$$

we obtain the homogeneous problem

$$Lu \equiv u'' - ku' = -k , \qquad u(0) = u(1) = 0 \tag{2.2b}$$

whose theoretical solution is

$$\phi = \frac{e^{kx} - e^k}{1 - e^k}$$

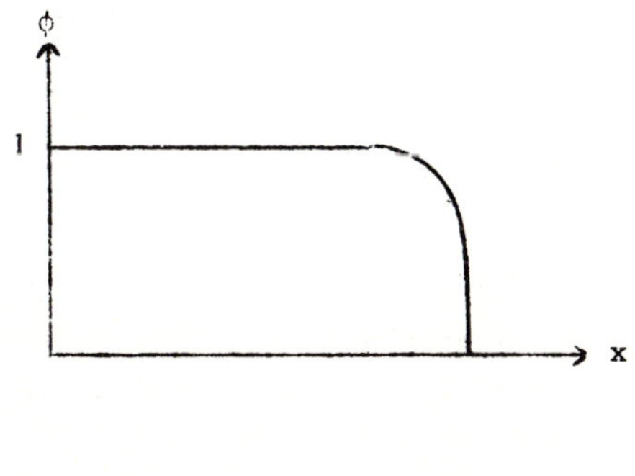

Figure 1.

If k is small, numerical solutions present no difficulty and when k is large numerical solutions of boundary layer type are available [6,7]. We are therefore interested in the middle range of values for k.

Associated with the operator $A \equiv \frac{d^2}{dx^2} - k\frac{d}{dx}$, we have the unsymmetric bilinear form

$$a(u,v) = -(u',v' + kv) , \tag{2.3}$$

and so the weak form of the original problem is:

find $u \in \overset{o}{H}{}^1[0,1]$ so that

$$(u',v' + kv) = (k,v) . \qquad \forall v \in \overset{o}{H}{}^1[0,1]. \tag{2.4}$$

We now let Π_h denote the subdivision of the interval $I = [0,1]$ into $(N+1)$ subintervals $I_j = [x_j, x_{j+1}]$, $j = 0,1,---,N$ with $x_0 = 0$, $x_{n+1} = 1$. Restricting attention to equal subdivision of I, we define $x_{j+1} - x_j = h$, $j = 0,1,---,N$. Associated with Π_h we have two finite dimensional approximations of $\overset{o}{H}{}^1[0,1]$ viz.

ϕ_h space of <u>trial</u> functions

ψ_h " " <u>test</u> "

and the discretised form of (2.4) becomes:

find $u_h \in \phi_h$ so that

$$a(u_h, v_h) = -(k, v_h) . \qquad \forall v_h \in \psi_h \tag{2.5}$$

3. Low Order Methods

If we use hat functions for both trial and test functions, we obtain the difference equation

$$(1-L)\phi_{j+1} - 2L\phi_j + (1+L)\phi_{j-1} = 0 , \qquad L = \tfrac{1}{2}hk$$

which is equivalent to using a central difference approximation for the first derivative. The theoretical solution of the difference equation contains the term

$$\left(\frac{1+L}{1-L}\right)^j \quad ((1,1) \text{ Padé approx to } e^{kx})$$

which oscillates with j if $h > \frac{2}{k}$. $(L > 1)$. As an example, if $k = 60$ and $h = 1/20$ $(L=3/2)$ the numerical results are

x	Theoretical	Approximate
0.90	0.9975	0.9600
0.95	0.9502	1.2000
1.00	0	0

showing considerable oscillations near $x = 1.0$. In order to cope with the unsymmetric nature of the significant first derivative term, it was decided to include asymmetry into the test functions. This was done in [8] by combining hat trial functions with test functions of the form

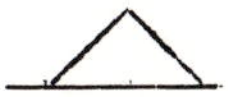 + an odd function with support 2h .

Two examples given in [8] are

(i)  $+ \alpha_1$ 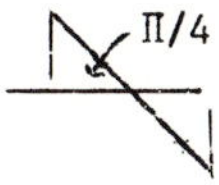

linear nonconforming

(ii)  $+ \alpha_2$  quadratics conforming

where $\alpha \geq 0$ $(i = 1,2)$. With the appropriate normalizations of the odd functions, the difference equations resulting from (i) and (ii) are

$$[1 - (1-\alpha_i)L]\phi_{j+1} - 2(1+\alpha_i L)\phi_j + [1 + (1+\alpha_i)L]\phi_{j-1} = 0 \,. \qquad (i=1,2)$$

The solutions to these equations contain no oscillations provided

(i) $\alpha_i \geq 1$

or

(ii) $-\infty < \alpha_i < 1$, $\qquad L < \dfrac{1}{1-\alpha_i}$.

Local accuracy is as follows:

(i)	first order	α_i = non-zero constant
(ii)	second order	$\alpha_i = 0$
(iii)	fourth order	$\alpha_i = \frac{1}{3}L$
(iv)	complete	$\alpha_i = (\coth L) - \frac{1}{L}$

In (iii), the theoretical solution of the difference equation contains the term

$$\left(\frac{1 + L + \frac{1}{3}L^2}{1 - L + \frac{1}{3}L^2}\right)^j \qquad ((2,2) \text{ Padé approximation to } e^{kh})$$

which is oscillation free for all L .

4. Higher order methods

(i) H^1 Galerkin method [9,10]. This is a generalised Galerkin method based on the use of the inner product in H^1 rather than in L_2. For problems in one dimension, the method is given by (1.1) with

$$\psi_i = \phi_i'' . \qquad i = 0,1,\text{---},N+1 .$$

For trial functions chosen as the Schoenberg cubic splines $B_i(x)$, the approximate solution U is given by

$$U = \sum_{i=0}^{N+1} \gamma_i B_i(x) ,$$

and the test functions are $B_i''(x)$, $i = 0,1,\text{---},N+1$. These functions are illustrated in Figure 2 and the trial and test functions at $Nh,(N+1)h$ are the mirror images of those at $h,0$. It is also shown in Figure 2(c) how each test function $B_i''(x)$ can be written as a linear combination of hat functions. Thus for trial spaces of Schoenberg cubic splines, the H^1 Galerkin method reduces to de Boor's method [11,12]. It should be noted that in the H^1 Galerkin method the trial functions satisfy the homogeneous boundary conditions but the test functions do <u>not</u>.

Applied to the homogeneous form of the model problem (2.2b) we obtain the set of equations

$$\begin{aligned}
&-(22+7L)\gamma_0 - (4+10L)\gamma_1 + (2-L)\gamma_2 &&= -\ 8hL\\
&-(4-6L)\gamma_0 - (14+L)\gamma_1 + (4-10L)\gamma_2 + (2-L)\gamma_3 &&= -16hL\\
&(2+L)\gamma_{i-2} + (4+10L)\gamma_{i-1} - 12\gamma_i + (4-10L)\gamma_{i+1} + (2-L)\gamma_{i+2} &&= -16hL\\
&\qquad (i=2,3,\text{---},N-1) && \qquad (4.1)\\
&(2+L)\gamma_{N-2} + (4+10L)\gamma_{N-1} - (14-L)\gamma_N - (4+6L)\gamma_{N+1} &&= -16hL\\
&(2+L)\gamma_{N-1} - (4-10L)\gamma_N - (22-7L)\gamma_{N+1} &&= -\ 8hL
\end{aligned}$$

where $L = \frac{1}{2}hk$. In order to study the problem of oscillations, we attempt a theoretical study of the difference system (4.1). The solution is given by

$$\gamma_i = \frac{2}{3} ih + A + B\lambda_1^{\,i} + C\lambda_2^{\,i} + D\lambda_3^{\,i}$$

where λ_1,λ_2 and λ_3 are the roots of the cubic

$$(2-L)\lambda^3 + (6-11L)\lambda^2 - (6+11L)\lambda - (2+L) = 0 \qquad (4.2)$$

and A,B,C and D are determined by the first and last pairs of equations in (4.1). However

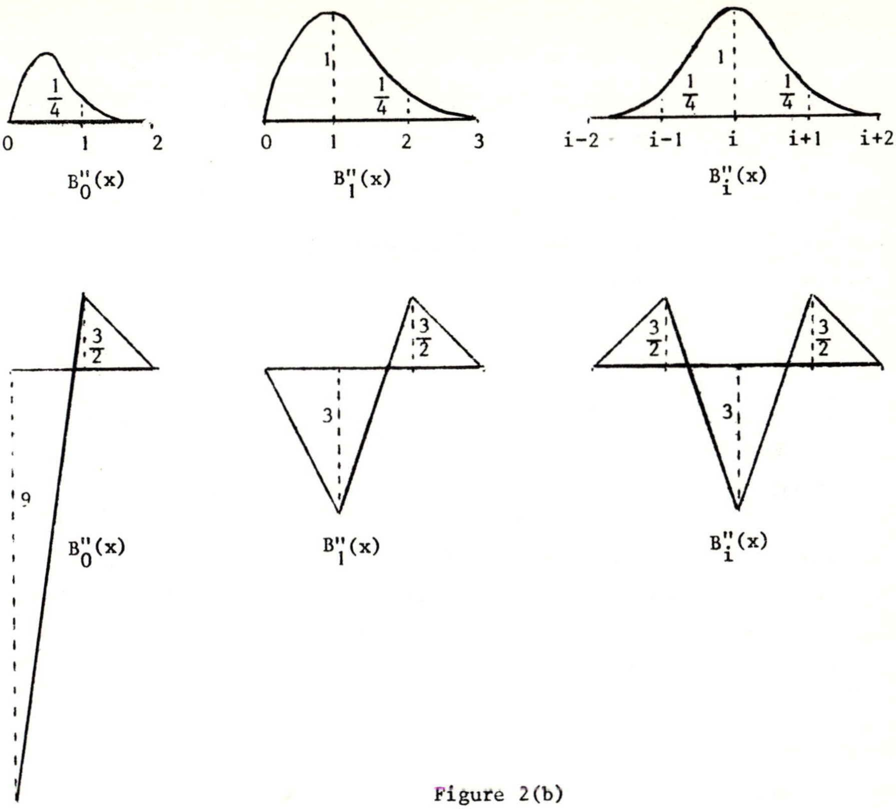

Figure 2(b)

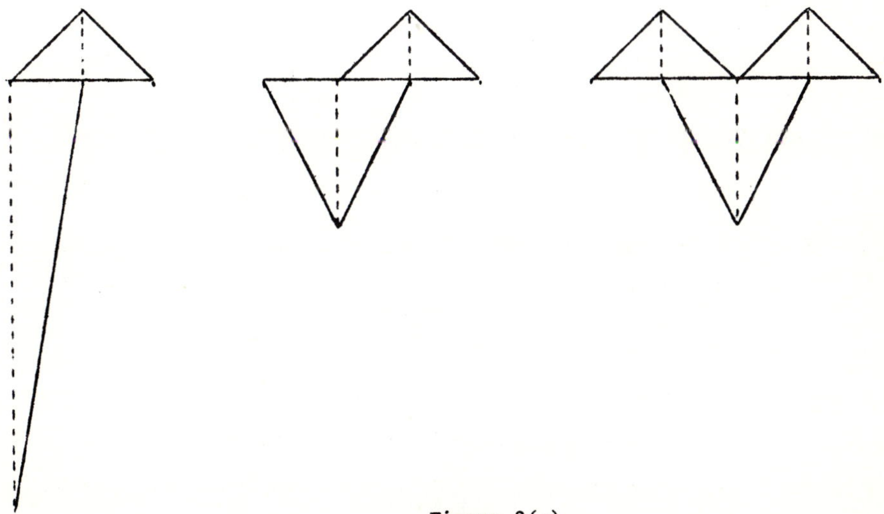

Figure 2(c)

$$U = \sum_{i=0}^{N+1} \gamma_i B_i(x) = \sum_{i=0}^{N+1} U_i C_i(x) ,$$

where $C_i(x)$ are the cardinal cubic splines, and so

$$U_0 = U_{N+1} = 0, \quad U_i = \frac{1}{4}\gamma_{i-1} + \gamma_i + \frac{1}{4}\gamma_{i+1} \qquad i = 1,2,---,N,$$

which leads to

$$U_i = ih + \frac{3}{2}A + \frac{1}{4}B\lambda_1^{\,i-1}(1+4\lambda_1+\lambda_1^2) + \frac{1}{4}C\lambda_2^{\,i-1}(1+4\lambda_2+\lambda_2^2) + \frac{1}{4}D\lambda_3^{\,i-1}(1+4\lambda_3+\lambda_3^2) \quad (4.3)$$

$$(i = 1,2,---,N).$$

Now as $L \to 0$, (4.2) factorises to give

$$(\lambda-1)(\lambda^2+4\lambda+1) = 0$$

and so the roots tend to $1, -(2 - \sqrt{3}), -(2 + \sqrt{3})$. The two negative real roots contribute significantly to the oscillations which appear in the numerical results in Table 1, but of course as $L \to 0$ the terms in (4.3) involving these roots disappear. Numerical results are shown in Table 1.

For trial functions which are <u>Hermite cubics</u>, $H(x)$, we consider the non homogeneous form of (2.2). The approximate solution Φ is given by

$$\Phi = \sum_{i=0}^{N+1} [\phi_i H_i(x) + \phi_i' \bar{H}_i(x)] ,$$

and the test functions are $H_i'(x)$ and $\bar{H}_i'(x)$, $i = 0,1,---,N+1$. These functions are illustrated in Figure 3. This time we obtain the system of equations

$$(6-2L)(\phi_0-\phi_1) + h(4+L)\phi_0' + h(2-L)\phi_1' = 0$$

$$\left.\begin{aligned} &6(\phi_{i-1}-2\phi_i+\phi_{i+1}) + h(3+L)\phi_{i-1}' - 2hL\phi_i' - h(3-L)\phi_{i+1}' = 0 \\ &(6+2L)\phi_{i-1} - 4L\phi_i - (6-2L)\phi_{i+1} + h(2+L)\phi_{i-1}' + 8h\phi_i' + h(2-L)\phi_{i+1}' = 0 \end{aligned}\right\} \; i=1,2,---,N \quad (4.4)$$

$$(6-2L)(\phi_{N+1}-\phi_N) - h(4+L)\phi_{N+1}' - h(2-L)\phi_N' = 0$$

where

$$\phi_0 = 1 , \qquad \phi_{N+1} = 0$$

from application of the boundary conditions. If we eliminate ϕ' from (4.4), the roots of the characteristic equation for ϕ are

$$1,\ 1,\ 1,\ \frac{1 + L + \frac{1}{3}L^2}{1 - L + \frac{1}{3}L^2}$$

The fourth root is positive for all $L(=\frac{1}{2}hk)$ and so ϕ is oscillation free. A similar argument based on eliminating ϕ from (4.4) leads to ϕ' being oscillation free. Note again that $\dfrac{1+L+\frac{1}{3}L^2}{1-L+\frac{1}{3}L^2}$ is the (2,2) Padé approximation to e^{kh}.

(ii) H^{-1} Galerkin method. [13,10] For two point boundary value problems with homogeneous boundary conditions we first define the spaces

$$M_r^s = \{v \mid v \in c^r(I),\ v \in P_s(I_j),\quad j = 0,1,\text{---},N\},\ r \geq 0$$

$$M_{-1}^s = \{v \mid v \in P_s(I_j),\quad j = 0,1,\text{---},N\}$$

$$N_r^s = M_{r+2}^{s+2} \cap \{v \mid v(0) = v(1) = 0\} .$$

The H^{-1} Galerkin procedure is:

let $U \in M_r^s$ satisfy

$$(U, A^*\phi) = (f,\phi) \qquad \forall\phi \in N_r^s \tag{4.5}$$

where A^* is the adjoint of the operator A. Note that it is ϕ, and not U, that is required to satisfy the boundary conditions. Examples of appropriate spaces are

(i) $r = 0$, $s = 1$.

$M_0^1 \equiv$ hat functions $\qquad N_0^1 \equiv$ cubic splines of figure 2a.

(ii) $r = -1$, $s = 1$.

$M_{-1}^1 \equiv$ discontinuous linears. $N_{-1}^1 \equiv$ cubic Hermites.

We give details of an H^{-1} Galerkin calculation involving the homogeneous form of the problem given by (2.2b) and the spaces in (i). The approximate solution U in (4.5) is given by

$$U = \sum_{i=0}^{N+1} U_i H_i(x) ,$$

where $H_i(x)$ is the hat function at node i. The linear system obtained is

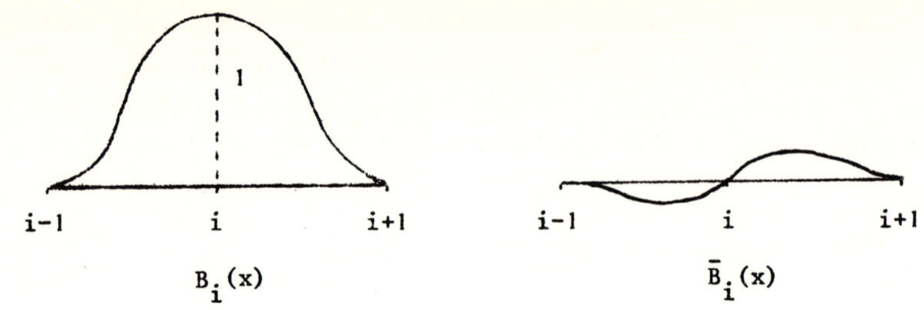

Figure 3(a)

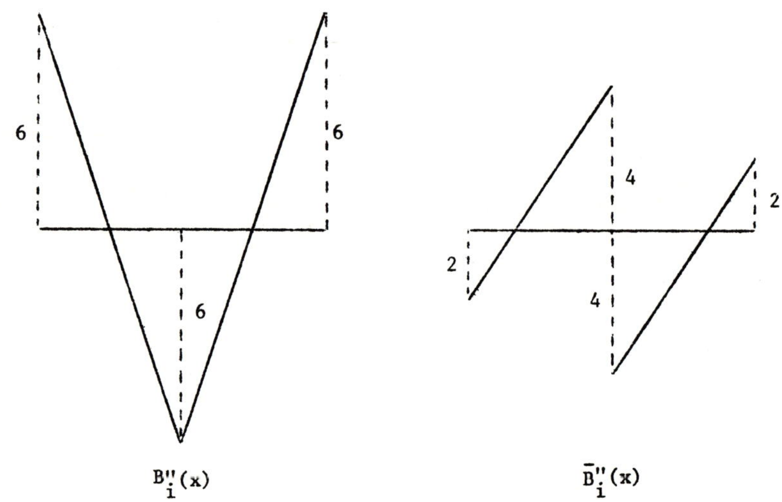

Figure 3(b)

$$-(22-7L)U_0 - (4+6L)U_1 + (2-L)U_2 = -8hL$$

$$-(4-10L)U_0 - (14-L)U_1 + (4-10L)U_2 + (2-L)U_3 = -22hL$$

$$(2+L)U_{i-2} + (4+10L)U_{i-1} - 12U_i + (4-10L)U_{i+1} + (2-L)U_{i+2} = -24hL$$

$$(i = 2,3,,---,N-1) \qquad (4.6)$$

$$(2+L)U_{N-2} + (4+10L)U_{N-1} - (14+L)U_N - (4+10L)U_{N+1} = -22hL$$

$$(2+L)U_{N-1} - (4-6L)U_N - (22+7L)U_{N+1} = -8hL$$

Numerical results are shown in Table 1.

(iii) Quadratic trial and test functions This is the conventional Galerkin method where the trial and test functions are given over the range $(i-1)h \leq x \leq ih$ $(i = 1,2,---,N+1)$ by

$$B_{i-1}(x) = \frac{1}{h^2}[2x^2 + (1-4i)hx + i(2i-1)h^2]$$

$$B_{i-\frac{1}{2}}(x) = \frac{1}{h^2}[-4x^2 + 4(2i-1)hx + 4i(1-i)h^2]$$

$$B_i(x) = \frac{1}{h^2}[2x^2 + (3-4i)hx + (2i-1)(i-1)h^2]$$

and sketched in Figure 4. The Galerkin solution leads

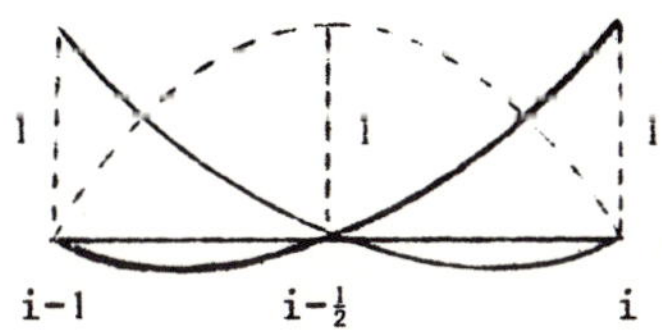

Figure 4

to the difference equations

$$(1-L)\phi_{i+1} - 4(2-L)\phi_{i+\frac{1}{2}} + 14\phi_i - 4(2+L)\phi_{i-\frac{1}{2}} + (1+L)\phi_{i-1} = 0 \qquad (i=1,2,---,N) \quad (4.7a)$$

at the integer nodes and

$$(1-\tfrac{1}{2}L)\phi_i - 2\phi_{i-\frac{1}{2}} + (1+\tfrac{1}{2}L)\phi_{i-1} = 0 \qquad (i = 1,2,---,N+1) \quad (4.7b)$$

at the half integer nodes. We can use (4.7b) to eliminate $\phi_{i-\frac{1}{2}}$ and $\phi_{i+\frac{1}{2}}$ in (4.7a). This leads to the result

$$(1-L+\tfrac{1}{3}L^2)\phi_{i+1} - 2(1+\tfrac{1}{3}L^2)\phi_i + (1+L+\tfrac{1}{3}L^2)\phi_{i-1} = 0 \qquad (i=1,2,---,N) \quad (4.8)$$

at the integer nodes with the theoretical solution

$$\phi_i = A + B\left(\frac{1 + L + \frac{1}{3}L^2}{1 - L + \frac{1}{3}L^2}\right)^i .$$

This solution is oscillation free for all L and $\dfrac{1 + L + \frac{1}{3}L^2}{1 - L + \frac{1}{3}L^2}$ is the (2,2) Padé approximation to e^{2L}. The theoretical solution at the half integer nodes is not the (2,2) Padé approximation and this may account for the contrasting behaviour of the solution at integer and half integer nodes which is often encountered in numerical calculations.

5. Two space dimensions

We now extend the lower order methods of section 3 to convective-diffusion problems in two dimensions and thereby present practical finite element applications. This work has already been published in [14]. It is assumed that the velocity field has been calculated in advance and that the direction of the velocity is known at every point in the field. If we use conforming elements, (section 3 (ii)), the signs of the weighting coefficient α_2 and the corresponding weighting coefficient in the y-direction, β_2 , are chosen to match respectively the directions of the velocity components u and v in (2.1) at the node in question. The magnitudes of the weighting coefficients are calculated from the formulae

$$\alpha_2 = \coth(\tfrac{1}{2}hu) - \frac{2}{hu} ,$$

and

$$\beta_2 = \coth(\tfrac{1}{2}hv) - \frac{2}{hv}$$

respectively and four problems of practical interest are successfully solved.

So far little progress has been made in generalising the higher order methods of section 4 to problems in two dimensions. The complicated H^1 and H^{-1} methods fail to produce satisfactory results in one dimensional problems. The promising Galerkin procedure with quadratic trial and test functions outlined in (iii) of section 4 does not appear to generalise to two dimensions. A recent procedure by Heinrich and Zienkiewicz [15] uses quadratic trial functions and cubic test functions but it does not appear to be significantly superior to the lower order method described in [14].

With regard to the nonlinear Navier Stokes equations where the velocity coefficients of the first order terms are not known in advance but are part of the solution itself, very little progress has been made for "reasonable" values of the Reynolds Number.

6. Time dependent problems

The transient form of convective-diffusion equations (of the form (2.1)), transport equations and the Navier Stokes equation when first derivatives in space are significant are even more difficult to analyse than the steady versions. The most convenient model problem takes the form

$$\frac{\partial u}{\partial t} = \frac{\partial^2 u}{\partial x^2} - k\frac{\partial u}{\partial x} \tag{6.1}$$

where $k(>0)$ is a constant, with the initial condition $u(x,0) = 0$ and the boundary conditions $u(0,t) = 1$ and $u(1,t) = 0$. In the nonlinear case, the simplest model is the Burgers equation

$$\frac{\partial u}{\partial t} = \nu \frac{\partial^2 u}{\partial x^2} - u \frac{\partial u}{\partial x} \tag{6.2}$$

where $\nu(>0)$ is a constant.

The principal methods of numerical solution of (6.1) discretise the space derivatives either by finite differences or by generalised Galerkin methods leading to a system of ordinary differential equations in time. The latter is solved by finite difference methods. Authors who have studied this problem numerically are Varga et al [16, 17, 18], Gladwell et al [19] and Heinrich and Zienkiewicz [15] . Oscillatory errors can now be produced by the time-discretisation (particularly in the early time steps) as well as by the space discretisation for significant values of k and the complete picture is one of confusion. Mass lumping [20] has been used to eliminate the oscillations in the early time steps, but this is less than satisfactory in many problems. Much work requires to be done in the transient case for problems involving second order equations with significant first derivatives.

Acknowledgement

The authors are indebted to Ian Christie, University of Dundee, for providing the numerical results in Table 1.

h = 1/10

	k = 30 (L = 1½)		k = 100 (L = 5)	
x	H^1	H^{-1}	H^1	H^{-1}
0	1.0000	.9991	1.0000	.9072
.1	.9994	1.0007	.7554	.9721
.2	1.0007	.9984	1.1912	.9138
.3	.9983	1.0025	.6425	.9822
.4	1.0026	.9950	1.3133	.8992
.5	.9949	1.0086	.4956	1.0003
.6	1.0087	.9841	1.4921	.8771
.7	.9837	1.0282	.2785	1.0273
.8	1.0267	.9476	1.7480	.8434
.9	.9001	1.0674	.0979	1.0805
1.0	0.0000	.2663	0.0000	.5964

h = 1/40

	k = 100 (L = 1¹/4)		k = 800 (L = 10)	
x	H^1	H^{-1}	H^1	H^{-1}
.75	1.0001	.9998	2.1776	.9521
.775	.9999	1.0003	- .3262	1.0278
.80	1.0002	.9994	2.4402	.9442
.825	.9995	1.0013	- .6162	1.0366
.85	1.0009	.9975	2.7606	.9345
.875	.9983	1.0048	- .9703	1.0473
.90	1.0033	.9905	3.1516	.9227
.925	.9929	1.0182	1.3999	1.0604
.95	1.0062	.9595	3.5990	.9074
.975	.8904	1.0198	-1.5725	1.0857
1.0	0.0000	.2175	0.0000	.7792

Table 1.

References

[1] Crandall, S H, "Engineering Analysis", McGraw Hill, New York. 1956.

[2] Collatz, L, "The numerical treatment of differential equations", Springer-Verlag, Berlin 1960.

[3] Anderssen, R S and Mitchell, A R, The Petrov-Galerkin method, Numerical Analysis Report 21. University of Dundee. 1977.

[4] Wachspress, E L, "A rational finite element basis", Academic Press, New York. 1975.

[5] Mitchell, A R and Wait, R, "The finite element method in partial differential equations", Wiley, London. 1977.

[6] Hemker, P W, "A numerical study of stiff two-point boundary problems", Mathematisch Centrum, Amsterdam. 1977.

[7] Griffiths, D F and Lorentz, J, "An analysis of the Petrov-Galerkin finite element method applied to a model problem", Research Paper 334, The University of Calgary. 1977.

[8] Christie, I, Griffiths, D F, Mitchell, A R and Zienkiewicz, O C, "Finite element methods for second order differential equations with significant first derivatives", Int. J. for Num. Meths. in Engng. 10, 1389-1396. 1976.

[9] Douglas, J, Dupont, T, and Wheeler, M F, "H^1-Galerkin methods for the Laplace and heat equations", Math. aspects of finite elements. ed. C de Boor, Academic Press, New York 1974.

[10] Lawlor, F M M, "The Galerkin method and its generalisations", M.Sc. Thesis, University of Dundee. 1976.

[11] de Boor, C R, "The method of projections as applied to the numerical solution of two point boundary value problems using cubic splines", Ph.D. Thesis, University of Michigan. 1966.

[12] Lucas, T R and Reddien, G W, "A high order projection method for nonlinear two point boundary value problems", Numer. Math. 20, 257-270. 1973.

[13] Rachford, H H and Wheeler, M F, "An H^{-1} Galerkin procedure for the two point boundary value problem", Maths aspects of finite elements, ed. C de Boor, Academic Press, New York. 1974.

[14] Heinrich, J C, Huyakorn, P S, Zienkiewicz, O C and Mitchell, A R, "An upwind finite element scheme for two dimensional convective transport equation", Int. J. for Num. Meths. in Engng, 11, 131-143. 1977.

[15] Heinrich, J C and Zienkiewicz, O C, "Quadratic finite element schemes for two dimensional convective-transport problems", Int. J. for Num. Meths. in Engng. (to appear).

[16] Price, H S, Varga, R S and Warren, J E, "Application of oscillation matrices to diffusion-convection equations", J. Math. Phys. 45, 1966.

[17] Price, H S, Cavendish, J C and Varga, R S, "Numerical methods of high-order accuracy for diffusion-convection equations", J. of Soc. Pet. Eng. 1963.

[18] Price, H S and Varga, R S, "Approximations of parabolic problems with applications to petroleum reservoir mechanics", SIAM AMS Proc 2, 1970.

[19] Siemieniuch, J L and Gladwell, I, "Some explicit finite-difference methods for the solution of a model diffusion-convection equation", Numerical Analysis Report 16, University of Manchester. 1976.

[20] Gresho, P M, Lee, R L and Sani, R L, "Advection dominated flows with emphasis on the consequence of mass lumping", ICCAD Second International Symposium on Finite Element Methods in Flow Problems, S Margherita Ligure Italy. 1976.

THE LEVENBERG-MARQUARDT ALGORITHM: IMPLEMENTATION AND THEORY*

Jorge J. Moré

1. Introduction

Let $F: R^n \to R^m$ be continuously differentiable, and consider the nonlinear least squares problem of finding a local minimizer of

$$\Phi(x) = \frac{1}{2} \sum_{i=1}^{m} f_i^2(x) = \frac{1}{2} \|F(x)\|^2 . \tag{1.1}$$

Levenberg [1944] and Marquardt [1963] proposed a very elegant algorithm for the numerical solution of (1.1). However, most implementations are either not robust, or do not have a solid theoretical justification. In this work we discuss a robust and efficient implementation of a version of the Levenberg-Marquardt algorithm, and show that it has strong convergence properties. In addition to robustness, the main features of this implementation are the proper use of implicitly scaled variables, and the choice of the Levenberg-Marquardt parameter via a scheme due to Hebden [1973]. Numerical results illustrating the behavior of this implementation are also presented.

Notation. In all cases $\|\cdot\|$ refers to the ℓ_2 vector norm or to the induced operator norm. The Jacobian matrix of F evaluated at x is denoted by $F'(x)$, but if we have a sequence of vectors $\{x_k\}$, then J_k and f_k are used instead of $F'(x_k)$ and $F(x_k)$, respectively.

2. Derivation

The easiest way to derive the Levenberg-Marquardt algorithm is by a linearization argument. If, given $x \in R^n$, we could minimize

$$\Psi(p) = \|F(x+p)\|$$

as a function of p, then x+p would be the desired solution. Since Ψ is usually a nonlinear function of p, we linearize F(x+p) and obtain the linear least squares problem

$$\psi(p) = \|F(x) + F'(x)p\| .$$

Of course, this linearization is not valid for all values of p, and thus we consider the constrained linear least squares problem

*Work performed under the auspices of the U.S. Energy Research and Development Administration

(2.1) $$\min\{\psi(p): \|Dp\| \le \Delta\} .$$

In theory D is any given nonsingular matrix, but in our implementation D is a diagonal matrix which takes into account the scaling of the problem. In either case, p lies in the hyperellipsoid

(2.2) $$E = \{p: \|Dp\| \le \Delta\} ,$$

but if D is diagonal, then E has axes along the coordinate directions and the length of the ith semi-axis is Δ/d_i.

We now consider the solution of (2.1) in some generality, and thus the problem

(2.3) $$\min\{\|f+Jp\|: \|Dp\| \le \Delta\}$$

where $f \in R^m$ and J is any m by n matrix. The basis for the Levenberg-Marquardt method is the result that if p^* is a solution to (2.3), then $p^* = p(\lambda)$ for some $\lambda \ge 0$ where

(2.4) $$p(\lambda) = -(J^TJ + \lambda D^TD)^{-1}J^Tf .$$

If J is rank deficient and $\lambda = 0$, then (2.4) is defined by the limiting process

$$Dp(0) \equiv \lim_{\lambda \to 0^+} Dp(\lambda) = -(JD^{-1})^{+}f .$$

There are two possibilities: Either $\lambda = 0$ and $\|Dp(0)\| \le \Delta$, in which case p(0) is the solution to (2.3) for which $\|Dp\|$ is least, or $\lambda > 0$ and $\|Dp(\lambda)\| = \Delta$, and then $p(\lambda)$ is the unique solution to (2.3).

The above results suggest the following iteration.

(2.5) Algorithm

(a) Given $\Delta_k > 0$, find $\lambda_k \ge 0$ such that if

$$(J_k^TJ_k + \lambda_kD_k^TD_k)p_k = -J_k^Tf_k ,$$

then either $\lambda_k = 0$ and $\|D_kp_k\| \le \Delta_k$, or $\lambda_k > 0$ and $\|D_kp_k\| = \Delta_k$.

(b) If $\|F(x_k+p_k)\| < \|F(x_k)\|$ set $x_{k+1} = x_k+p_k$ and evaluate J_{k+1}; otherwise set $x_{k+1} = x_k$ and $J_{k+1} = J_k$.

(c) Choose Δ_{k+1} and D_{k+1}.

In the next four sections we elaborate on how (2.5) leads to a very robust and efficient implementation of the Levenberg-Marquardt algorithm.

3. Solution of a Structured Linear Least Squares Problem

The simplest way to obtain the correction p is to use Cholesky decomposition on the linear system

$$(3.1) \qquad (J^TJ + \lambda D^TD)p = -J^Tf \ .$$

Another method is to recognize that (3.1) are the normal equations for the least squares problem

$$(3.2) \qquad \begin{pmatrix} J \\ \lambda^{\frac{1}{2}}D \end{pmatrix} p \cong - \begin{pmatrix} f \\ 0 \end{pmatrix} ,$$

and to solve this structured least squares problem using QR decomposition with column pivoting.

The main advantage of the normal equations is speed; it is possible to solve (3.1) twice as fast as (3.2). On the other hand, the normal equations are particularly unreliable when $\lambda = 0$ and J is nearly rank deficient. Moreover, the formation of J^TJ or D^TD can lead to unnecessary underflows and overflows, while this is not the case with (3.2). We feel that the loss in speed is more than made up by the gain in reliability and robustness.

The least squares solution of (3.2) proceeds in two stages. These stages are the same as those suggested by Golub (Osborne [1972]), but modified to take into account the pivoting.

In the first stage, compute the QR decomposition of J with column pivoting. This produces an orthogonal matrix Q and a permutation π of the columns of J such that

$$(3.3) \qquad QJ\pi = \begin{pmatrix} T & S \\ 0 & 0 \end{pmatrix}$$

where T is a nonsingular upper triangular matrix of rank (J) order. If $\lambda = 0$, then a solution of (3.2) is

$$p = \pi \begin{pmatrix} T^{-1} & 0 \\ 0 & 0 \end{pmatrix} Qf \equiv J^-f$$

where J^- refers to a particular symmetric generalized inverse of J in the sense that JJ^- is symmetric and $JJ^-J = J$. To solve (3.2) when $\lambda > 0$ first note that (3.3) implies that

$$(3.4) \qquad \begin{pmatrix} Q & 0 \\ 0 & \pi^T \end{pmatrix} \begin{pmatrix} J \\ \lambda^{\frac{1}{2}}D \end{pmatrix} \pi = \begin{pmatrix} R \\ 0 \\ D_\lambda \end{pmatrix}$$

where $D_\lambda = \lambda^{\frac{1}{2}}\pi^TD\pi$ is still a diagonal matrix and R is a (possibly singular) upper triangular matrix of order n.

In the second stage, compute the QR decomposition of the matrix on the right of (3.4). This can be done with a sequence of $n(n+1)/2$ Givens rotations. The result is an orthogonal matrix W such that

(3.5)
$$W\begin{pmatrix} R \\ 0 \\ D_\lambda \end{pmatrix} = \begin{pmatrix} R_\lambda \\ 0 \end{pmatrix}$$

where R_λ is a nonsingular upper triangular matrix of order n. The solution to (3.2) is then

$$p = -\pi R_\lambda^{-1} u$$

where $u \in R^n$ is determined from

$$W\begin{pmatrix} Qf \\ 0 \end{pmatrix} = \begin{pmatrix} u \\ v \end{pmatrix} .$$

It is important to note that if λ is changed, then only the second stage must be redone.

4. Updating the Step Bound

The choice of Δ depends on the ratio between the actual reduction and the predicted reduction obtained by the correction. In our case, this ratio is given by

(4.1)
$$\rho(p) = \frac{\|F(x)\|^2 - \|F(x+p)\|^2}{\|F(x)\|^2 - \|F(x)+F'(x)p\|^2} .$$

Thus (4.1) measures the agreement between the linear model and the (nonlinear) function. For example, if F is linear then $\rho(p) = 1$ for all p, and if $F'(x)^T F(x) \neq 0$, then $\rho(p) \to 1$ as $\|p\| \to 0$. Moreover, if $\|F(x+p)\| \geq \|F(x)\|$ then $\rho(p) \leq 0$.

The scheme for updating Δ has the objective of keeping the value of (4.1) at a reasonable level. Thus, if $\rho(p)$ is close to unity (i.e. $\rho(p) \geq 3/4$), we may want to increase Δ, but if $\rho(p)$ is not close to unity (i.e. $\rho(p) \leq 1/4$), then Δ must be decreased. Before giving more specific rules for updating Δ, we discuss the computation of (4.1). For this, write

(4.2)
$$\rho = \frac{\|f\|^2 - \|f_+\|^2}{\|f\|^2 - \|f+Jp\|^2}$$

with an obvious change in notation. Since p satisfies (3.1),

(4.3)
$$\|f\|^2 - \|f+Jp\|^2 = \|Jp\|^2 + 2\lambda\|Dp\|^2 ,$$

and hence we can rewrite (4.2) as

$$(4.4) \qquad \rho = \frac{1 - \left(\frac{\|f_+\|}{\|f\|}\right)^2}{\left(\frac{\|Jp\|}{\|f\|}\right)^2 + 2\left(\frac{\lambda^{\frac{1}{2}}\|Dp\|}{\|f\|}\right)^2} .$$

Since (4.3) implies that

$$\|Jp\| \le \|f\|, \quad \lambda^{\frac{1}{2}}\|Dp\| \le \|f\| ,$$

the computation of the denominator will not generate any overflows, and moreover, the denominator will be non-negative regardless of roundoff errors. Note that this is not the case with (4.2). The numerator of (4.4) may generate overflows if $\|f_+\|$ is much larger than $\|f\|$, but since we are only interested in positive values of ρ, if $\|f_+\| > \|f\|$ we can just set $\rho = 0$ and avoid (4.4).

We now discuss how to update Δ. To increase Δ we simply multiply Δ by a constant factor not less than one. To decrease Δ we follow Fletcher [1971] and fit a quadratic to $\delta(0)$, $\delta'(0)$ and $\delta(1)$ where

$$\delta(\theta) = \frac{1}{2}\|F(x+\theta p)\|^2 .$$

If μ is the minimizer of the resulting quadratic, we decrease Δ by multiplying Δ by μ, but if $\mu \notin \left(\frac{1}{10}, \frac{1}{2}\right)$, we replace μ by the closest endpoint. To compute μ safely, first note that (3.1) implies that

$$\gamma \equiv \frac{p^TJ^Tf}{\|f\|^2} = -\left[\left(\frac{\|Jp\|}{\|f\|}\right)^2 + \left(\lambda^{\frac{1}{2}}\frac{\|Dp\|}{\|f\|}\right)^2\right] ,$$

and that $\gamma \in [-1,0]$. It is now easy to verify that

$$(4.5) \qquad \mu = \frac{\frac{1}{2}\gamma}{\gamma + \frac{1}{2}\left[1 - \left(\frac{\|f_+\|}{\|f\|}\right)^2\right]} .$$

If $\|f_+\| \le \|f\|$ we set $\mu = 1/2$. Also note that we only compute μ by (4.5) if say, $\|f_+\| \le 10\|f\|$, for otherwise, $\mu \le 1/10$.

5. The Levenberg-Marquardt Parameter

In our implementation $\alpha > 0$ is accepted as the Levenberg-Marquardt parameter if

$$(5.1) \qquad |\phi(\alpha)| \le \sigma\Delta ,$$

where

$$(5.2) \qquad \phi(\alpha) = \|D(J^TJ+\alpha D^TD)^{-1}J^Tf\| - \Delta ,$$

and $\sigma \in (0,1)$ specifies the desired relative error in $\|Dp(\alpha)\|$. Of course, if

$\phi(0) \leq 0$ then $\alpha = 0$ is the required parameter, so in the remainder of this section we assume that $\phi(0) > 0$. Then ϕ is a continuous, strictly decreasing function on $[0,+\infty)$ and $\phi(\alpha)$ approaches $-\Delta$ at infinity. It follows that there is a unique $\alpha^* > 0$ such that $\phi(\alpha^*) = 0$. To determine the Levenberg-Marquardt parameter we assume that an initial estimate $\alpha_0 > 0$ is available, and generate a sequence $\{\alpha_k\}$ which converges to α^*.

Since ϕ is a convex function, it is very tempting to use Newton's method to generate $\{\alpha_k\}$, but this turns out to be very inefficient -- the particular structure of this problem allows us to derive a much more efficient iteration due to Hebden [1973]. To do this, note that

$$(5.3)\qquad \phi(\alpha) = \| (\tilde{J}^T\tilde{J}+\alpha I)^{-1}\tilde{J}^T f \| - \Delta, \quad \tilde{J} = JD^{-1} ,$$

and let $\tilde{J} = U\Sigma V^T$ be the singular value decomposition of $\tilde{J}$. Then

$$\phi(\alpha) = \left[\sum_{i=1}^{n} \frac{\sigma_i^2 z_i^2}{(\sigma_i^2+\alpha)^2} \right]^{\frac{1}{2}} - \Delta$$

where $z = U^T f$ and $\sigma_1,\ldots,\sigma_n$ are the singular values of $\tilde{J}$. Hence, it is very natural to assume that

$$\phi(\alpha) \doteq \frac{a}{b + \alpha} - \Delta \equiv \tilde{\phi}(\alpha) ,$$

and to choose a and b so that $\tilde{\phi}(\alpha_k) = \phi(\alpha_k)$ and $\tilde{\phi}'(\alpha_k) = \phi'(\alpha_k)$. Then $\tilde{\phi}(\alpha_{k+1}) = 0$ if

$$(5.4)\qquad \alpha_{k+1} - \alpha_k \left[\frac{\phi(\alpha_k) + \Delta}{\Delta}\right] \left[\frac{\phi(\alpha_k)}{\phi'(\alpha_k)}\right] .$$

This iterative scheme must be safeguarded if it is to converge. Hebden [1973] proposed using upper and lower bounds u_k and ℓ_k, and that (5.4) be applied with the restriction that no iterate may be within $(u_k-\ell_k)/10$ of either endpoint. It turns out that this restriction is very detrimental to the progress of the iteration since in a lot of cases u_k is much larger than ℓ_k. A much more efficient algorithm can be obtained if (5.4) is only modified when α_{k+1} is outside of (ℓ_{k+1},u_{k+1}). To specify this algorithm we first follow Hebden [1973] and note that (5.3) implies that

$$u_0 = \frac{\| (JD^{-1})^T f \|}{\Delta}$$

is a suitable upper bound. If J is not rank deficient, then $\phi'(0)$ is defined and the convexity of ϕ implies that

$$\ell_0 = - \frac{\phi(0)}{\phi'(0)}$$

is a lower bound; otherwise let $\ell_0 = 0$.

(5.5) Algorithm

(a) If $\alpha_k \notin (\ell_k, u_k)$ let $\alpha_k = \max\{0.001\, u_k, (\ell_k u_k)^{1/2}\}$.

(b) Evaluate $\phi(\alpha_k)$ and $\phi'(\alpha_k)$. Update u_k by letting $u_{k+1} = \alpha_k$ if $\phi(\alpha_k) < 0$ and $u_{k+1} = u_k$ otherwise. Update ℓ_k by

$$\ell_{k+1} = \max\left\{\ell_k, \alpha_k - \frac{\phi(\alpha_k)}{\phi'(\alpha_k)}\right\} .$$

(c) Obtain α_{k+1} from (5.4).

The role of (5.5)(a) is to replace α_k by a point in (ℓ_k, u_k) which is biased towards ℓ_k; the factor $0.001\, u_k$ was added to guard against exceedingly small values of ℓ_k, and in particular, $\ell_k = 0$. In (5.5)(b), the convexity of ϕ guarantees that the Newton iterate can be used to update ℓ_k.

It is not too difficult to show that algorithm (5.5) always generates a sequence which converges quadratically to α^*. In practice, less than two iterations (on the average) are required to satisfy (5.1) when $\sigma = 0.1$.

To complete the discussion of the Hebden algorithm, we show how to evaluate $\phi'(\alpha)$. From (5.2) it follows that

$$\phi'(\alpha) = -\frac{(D^Tq(\alpha))^T(J^TJ+\alpha D^TD)^{-1}(D^Tq(\alpha))}{\|q(\alpha)\|}$$

where $q(\alpha) = Dp(\alpha)$ and $p(\cdot)$ is defined by (2.4). From (3.4) and (3.5) we have

$$\pi^T(J^TJ+\alpha D^TD)\pi = R_\alpha^T R_\alpha ,$$

and hence,

$$\phi'(\alpha) = -\|q(\alpha)\| \left\| R_\alpha^{-T}\left(\frac{\pi^TD^Tq(\alpha)}{\|q(\alpha)\|}\right)\right\|^2 .$$

6. Scaling

Since the purpose of the matrix D_k in the Levenberg-Marquardt algorithm is to take into account the scaling of the problem, some authors (e.g. Fletcher [1971]) choose

(6.1) $$D_k = \mathrm{diag}(d_1^{(k)}, \ldots, d_n^{(k)})$$

where

(6.2) $$d_i^{(k)} = \|\partial_i F(x_0)\|, \quad k \geq 0 .$$

This choice is usually adequate as long as $\|\partial_i F(x_k)\|$ does not increase with k. However, if $\|\partial_i F(x_k)\|$ increases, this requires a decrease in the length ($= \Delta/d_i$) of the ith semi-axis of the hyperellipsoid (2.2), since F is now changing faster along the

i^{th} variable, and therefore, steps which have a large i^{th} component tend to be unreliable. This argument leads to the choice

$$d_i^{(0)} = \|\partial_i F(x_0)\| \tag{6.3}$$

$$d_i^{(k)} = \max\left\{d_i^{(k-1)}, \|\partial_i F(x_k)\|\right\} , \quad k \geq 1 .$$

Note that a decrease in $\|\partial_i F(x_k)\|$ only implies that F is not changing as fast along the ith variable, and hence does not require a decrease in d_i. In fact, the choice

$$d_i^{(k)} = \|\partial_i F(x_k)\| , \quad k \geq 0 , \tag{6.4}$$

is computationally inferior to both (6.2) and (6.3). Moreover, our theoretical results support choice (6.3) over (6.4), and to a lesser extent, (6.2).

It is interesting to note that (6.2), (6.3), and (6.4) make the Levenberg-Marquardt algorithm scale invariant. In other words, for all of the above choices, if D is a diagonal matrix with positive diagonal elements, then algorithm (2.5) generates the same iterates if either it is applied to F and started at x_0, or if it is applied to $\tilde{F}(x) = F(D^{-1}x)$ and started at $\tilde{x}_0 = Dx_0$. For this result it is assumed that the decision to change Δ is only based on (4.1), and thus is also scale invariant.

7. Theoretical Results

It will be sufficient to present a convergence result for the following version of the Levenberg-Marquardt algorithm.

(7.1) Algorithm

(a) Let $\sigma \in (0,1)$. If $\|D_k J_k^- f_k\| \leq (1+\sigma)\Delta_k$, set $\lambda_k = 0$ and $p_k = -J_k^- f_k$. Otherwise determine $\lambda_k > 0$ such that if

$$\begin{bmatrix} J_k \\ \lambda_k^{\frac{1}{2}} D_k \end{bmatrix} p_k \cong - \begin{bmatrix} f_k \\ 0 \end{bmatrix}$$

then

$$(1-\sigma)\Delta_k \leq \|D_k p_k\| \leq (1+\sigma)\Delta_k .$$

(b) Compute the ratio ρ_k of actual to predicted reduction.

(c) If $\rho_k \leq 0.0001$, set $x_{k+1} = x_k$ and $J_{k+1} = J_k$.
If $\rho_k > 0.0001$, set $x_{k+1} = x_k + p_k$ and compute J_{k+1}.

(d) If $\rho_k \leq 1/4$, set $\Delta_{k+1} \in \left[\frac{1}{10}\Delta_k, \frac{1}{2}\Delta_k\right]$.
If $\rho_k \in \left(\frac{1}{4}, \frac{3}{4}\right)$ and $\lambda_k = 0$, or if $\rho_k \geq 3/4$, set $\Delta_{k+1} = 2\|D_k p_k\|$.

(e) Update D_{k+1} by (6.1) and (6.3).

The proof of our convergence result is somewhat long and will therefore be presented elsewhere.

Theorem. Let $F: R^n \to R^m$ be continuously differentiable on R^n, and let $\{x_k\}$ be the sequence generated by algorithm (7.1). Then

$$\liminf_{k \to +\infty} \|(J_k D_k^{-1})^T f_k\| = 0 . \tag{7.2}$$

This result guarantees that eventually a scaled gradient will be small enough. Of course, if $\{J_k\}$ is bounded then (7.2) implies the more standard result that

$$\liminf_{k \to +\infty} \|J_k^T f_k\| = 0 . \tag{7.3}$$

Furthermore, we can also show that if F' is uniformly continuous then

$$\lim_{k \to +\infty} \|J_k^T f_k\| = 0 . \tag{7.4}$$

Powell [1975] and Osborne [1975] have also obtained global convergence results for their versions of the Levenberg-Marquardt algorithm. Powell presented a general algorithm for unconstrained minimization which as a special case contains (7.1) with $\sigma = 0$ and $\{D_k\}$ constant. For this case Powell obtains (7.3) under the assumption that $\{J_k\}$ is bounded. Osborne's algorithm directly controls $\{\lambda_k\}$ instead of $\{\Delta_k\}$, and allows $\{D_k\}$ to be chosen by (6.1) and (6.3). For this case he proves (7.4) under the assumptions that $\{J_k\}$ and $\{\lambda_k\}$ are bounded.

8. Numerical Results

In our numerical results we would like to illustrate the behavior of our algorithm with the three choices of scaling mentioned in Section 6. For this purpose, we have chosen four functions.

1) Fletcher and Powell [1963] n=3, m=3

$$f_1(x) = 10[x_3 - 10\theta(x_1,x_2)]$$
$$f_2(x) = 10[(x_1^2+x_2^2)^{\frac{1}{2}} - 1]$$
$$f_3(x) = x_3$$

where

$$\theta(x_1,x_2) = \begin{cases} \frac{1}{2\pi} \arctan (x_2/x_1), & x_1 > 0 \\ \frac{1}{2\pi} \arctan (x_2/x_1) + 0.5, & x_1 < 0 \end{cases}$$

$$x_0 = (-1,0,0)^T$$

2. Kowalik and Osborne [1968] $n=4$, $m=11$

$$f_i(x) = y_i - \frac{x_1[u_i^2 + x_2u_i]}{(u_i^2 + x_3u_i + x_4)}$$

where u_i and y_i are specified in the original paper.

$$x_0 = (0.25, 0.39, 0.415, 0.39)^T$$

3. Bard [1970] $n=3$, $m=15$

$$f_i(x) = y_i - \left[x_1 + \frac{u_i}{x_2v_i + x_3w_i}\right]$$

where $u_i = i$, $v_i = 16-i$, $w_i = \min\{u_i,v_i\}$, and y_i is specified in the original paper.

$$x_0 = (1,1,1)^T$$

4. Brown and Dennis [1971] $n=4$, $m=20$

$$f_i(x) = [x_1 + x_2t_i - \exp(t_i)]^2 + [x_3 + x_4\sin(t_i) - \cos(t_i)]^2$$

where $t_i = (0.2)i$.

$$x_0 = (25, 5, -5, 1)^T$$

These problems have very interesting features. Problem 1 is a helix with a zero residual at $x^* = (1,0,0)$ and a discontinuity along the plane $x_1 = 0$; note that the algorithm must cross this plane to reach the solution. Problems 2 and 3 are data fitting problems with small residuals, while Problem 4 has a large residual. The residuals are given below.

1. $\|F(x^*)\| = 0.0$
2. $\|F(x^*)\| = 0.0175358$
3. $\|F(x^*)\| = 0.0906359$
4. $\|F(x^*)\| = 292.9542$

Problems 2 and 3 have other solutions. To see this, note that for Kowalik and Osborne's function,

$$\lim_{\alpha\to\infty} f_i(\alpha,x_2,\alpha,\alpha) = y_i - \left(\frac{u_i}{u_i+1}\right)(x_2+u_i) , \tag{8.1}$$

while for Bard's function,

$$\lim_{\alpha\to\infty} f_i(x_1,\alpha,\alpha) = y_i - x_1 . \tag{8.2}$$

These are now linear least squares problems, and as such, the parameter x_2 in (8.1) and x_1 in (8.2) are completely determined. However, the remaining parameters only need to be sufficiently large.

In presenting numerical results one must be very careful about the convergence criteria used. This is particularly true of the Levenberg-Marquardt method since, unless $F(x^*) = 0$, the algorithm converges linearly. In our implementation, an approximation x to x^* is acceptable if either x is close to x^* or $\|F(x)\|$ is close to $\|F(x^*)\|$. We attempt to satisfy these criteria by the convergence tests

$$\Delta \leq \text{XTOL}\ \|Dx\| , \tag{8.3}$$

and

$$\left(\frac{\|Jp\|}{\|f\|}\right)^2 + 2\left(\lambda^{\frac{1}{2}}\frac{\|Dp\|}{\|f\|}\right)^2 \leq \text{FTOL} . \tag{8.4}$$

An important aspect of these tests is that they are scale invariant in the sense of Section 6. Also note that the work of Section 4 shows that (8.4) is just the relative error between $\|f+Jp\|^2$ and $\|f\|^2$.

The problems were run on the IBM 370/195 of Argonne National Laboratory in double precision (14 hexadecimal digits) and under the FORTRAN H (opt=2) compiler. The tolerances in (8.3) and (8.4) were set at FTOL = 10^{-8} and XTOL = 10^{-8}. Each problem is run with three starting vectors. We have already given the starting vector x_0 which is closest to the solution; the other two points are $10x_0$ and $100x_0$. For each starting vector, we have tried our algorithm with the three choices of $\{D_k\}$. In the table below, choices (6.2), (6.3) and (6.4) are referred to as initial, adaptive, and continuous scaling, respectively. Moreover, NF and NJ stands for the number of function and Jacobian evaluations required for convergence.

		x_0		$10x_0$		$100x_0$	
PROBLEM	SCALING	NF	NJ	NF	NJ	NF	NJ
1	Initial	12	9	34	29	FC	FC
	Adaptive	11	8	20	15	19	16
	Continuous	12	9	14	12	176	141
2	Initial	19	17	81	71	365	315
	Adaptive	18	16	79	71	348	307
	Continuous	18	16	63	54	FC	FC
3	Initial	8	7	37	36	14	13
	Adaptive	8	7	37	36	14	13
	Continuous	8	7	FC	FC	FC	FC
4	Initial	268	242	423	400	FC	FC
	Adaptive	268	242	57	47	229	207
	Continuous	FC	FC	FC	FC	FC	FC

Interestingly enough, convergence to the minimizer indicated by (8.1) only occurred for starting vector $10x_0$ of Problem 2, while for Problem 3 starting vectors $10x_0$ and $100x_0$ led to (8.2). Otherwise, either the global minimizer was obtained, or the algorithm failed to converge to a solution; the latter is indicated by FC in the table.

It is clear from the table that the adaptive strategy is best in these four examples. We have run other problems, but in all other cases the difference is not as dramatic as in these cases. However, we believe that the above examples adequately justify our choice of scaling matrix.

Acknowledgments. This work benefited from interaction with several people. Beverly Arnoldy provided the numerical results for several versions of the Levenberg-Marquardt algorithm, Brian Smith showed how to use pivoting in the two-stage process of Section 3, and Danny Sorensen made many valuable comments on an earlier draft of this paper. Finally, I would like to thank Judy Beumer for her swift and beautiful typing of the paper.

References

1. Bard, Y. [1970]. Comparison of gradient methods for the solution of nonlinear parameter estimation problem, SIAM J. Numer. Anal. 7, 157-186.

2. Brown, K. M. and Dennis, J. E. [1971]. New computational algorithms for minimizing a sum of squares of nonlinear functions, Department of Computer Science report 71-6, Yale University, New Haven, Connecticut.

3. Fletcher, R. [1971]. A modified Marquardt subroutine for nonlinear least squares, Atomic Energy Research Establishment report R6799, Harwell, England.

4. Fletcher, R. and Powell, M.J.D. [1963]. A rapidly convergent descent method for minimization, Comput. J. 6, 163-168.

5. Hebden, M. D. [1973]. An algorithm for minimization using exact second derivatives, Atomic Energy Research Establishment report TP515, Harwell, England.

6. Kowalik, J. and Osborne, M. R. [1968]. Methods for Unconstrained Optimization Problems, American Elsevier.

7. Levenberg, K. [1944]. A method for the solution of certain nonlinear problems in least squares, Quart. Appl. Math. 2, 164-168.

8. Marquardt, D. W. [1963]. An algorithm for least squares estimation of nonlinear parameters, SIAM J. Appl. Math. 11, 431-441.

9. Osborne, M. R. [1972]. Some aspects of nonlinear least squares calculations, in Numerical Methods for Nonlinear Optimization, F. A. Lootsma, ed., Academic Press.

10. Osborne, M. R. [1975]. Nonlinear least squares - the Levenberg algorithm revisited, to appear in Series B of the Journal of the Australian Mathematical Society.

11. Powell, M. J. D. [1975]. Convergence properties of a class of minimization algorithms, in Nonlinear Programming 2, O. L. Mangasarian, R. R. Meyer, and S. M. Robinson, eds., Academic Press.

Nonlinear Approximation Problems in Vector Norms

M R Osborne and G A Watson

1. Introduction

Let $\underset{\sim}{f}(\underset{\sim}{x}) : R^n \to R^m$, where $n < m$, be a vector valued function, in general nonlinear in its dependence on its argument vector $\underset{\sim}{x}$. The class of problems we consider can be formulated as

$$\text{find } \underset{\sim}{x} \in R^n \text{ to minimise } \|\underset{\sim}{f}\| \tag{1.1}$$

where $\|\cdot\|$ is some norm defined on R^m. Problems of this kind occur for example in discrete data analysis and parameter estimation problems. The norm is characteristically the L_2 norm in which case (1.1) provides the maximum likelihood estimator appropriate to the normal probability distribution. However, use of the L_1 norm provides a parameter estimation procedure which is typically less sensitive to the effects of blunders or gross errors in data collection, and so has proved attractive as a basis for robust estimation procedures. Also the L_∞ norm is relevant to minimising the absolute error in the approximate representation of functions and thus finds application in providing approximations to functions suitable for evaluation by computer subroutines.

Because $\underset{\sim}{f}(\underset{\sim}{x})$ is assumed nonlinear it is, in general, necessary to employ an iterative procedure for the solution of (1.1). The Gauss-Newton algorithm is perhaps most used when the norm is the L_2 norm, and this has been generalised both to the L_∞ norm and the L_1 norm in Osborne and Watson (1969) and (1971). The form of generalisation is quite straightforward. A linear approximation to $\underset{\sim}{f}(\underset{\sim}{x})$ is minimised in the appropriate norm to determine a direction $\underset{\sim}{h}$ (this is called the linear subproblem or LSP), then this direction is used as the basis for a descent step for reducing $\|\underset{\sim}{f}\|$ by means of a suitable linear search strategy. The point reached is taken as the next approximation, and the whole procedure repeated iteratively. Osborne (1972) noted for the particular cases of L_2 and L_∞ that the properties of the iteration depend on the norm used. In particular the perhaps surprising result that the L_∞ norm iteration could have a faster rate of convergence than was the case in the L_2 norm was indicated. Thus it is not apparent that the L_2 norm is the appropriate choice if the requirement is just to make the components of $\underset{\sim}{f}$ small.

Our aim in this paper is to extend these results to more general norms. In section 2, we establish conditions for stationary points of $\|\underset{\sim}{f}\|$, and we give a generalisation of the Gauss-Newton algorithm in Section 3, showing that the limit points of the iteration are stationary points of $\|\underset{\sim}{f}\|$ provided that $\nabla\underset{\sim}{f}(\underset{\sim}{x})$ has full rank. Rate of convergence results are possible if the full step method is convergent, and conditions which achieve this are given in section 4. In particular, it turns out to be important to obtain certain bounds on the solution $\underset{\sim}{h}$ to the LSP, and it is here that the differences between the norms becomes apparent. Results for polyhedral norms (which include the L_1 and L_∞ cases) have been given by Anderson and Osborne (1977a) and these are summarised in section 5. The case of smooth strictly convex, monotonic norms is considered in section 6. The requirement for $\nabla \underset{\sim}{f}$ to be of full rank is a characteristic sufficiency condition associated with the Gauss-Newton algorithm, and certainly the method does not appear satisfactory in other cases, even if it need not actually fail. For this reason, modifications of the LSP to overcome problems associated with rank deficiency are important. For nonlinear least squares, this includes the ubiquitous Levenberg algorithm, and generalisations to L_∞ (Madsen (1975)) and to polyhedral norms (Anderson and Osborne (1977b)) have been given. It is possible to give a general formulation embracing all these particular cases, and the basic norm properties required are summarised in section 7.

For the purposes of our presentation, we assume that there exists a stationary point of $\|\underset{\sim}{f}\|$ in some bounded region S, and that the algorithm produces iterates in S. Essentially we assume that a local analysis suffices, so that any necessary properties of smoothness, boundedness etc. hold. In particular, we will assume that $\underset{\sim}{f}$ is sufficiently smooth that

$$\underset{\sim}{f}(\underset{\sim}{x}+\underset{\sim}{t}) = \underset{\sim}{f}(\underset{\sim}{x}) + A\underset{\sim}{t} + \|\underset{\sim}{t}\|_H^2 \; \underset{\sim}{w}(\underset{\sim}{x},\underset{\sim}{t}) \tag{1.2}$$

where $\underset{\sim}{x}, \underset{\sim}{x}+\underset{\sim}{t} \in S$, $A = \nabla\underset{\sim}{f}(\underset{\sim}{x})$, $\|\underset{\sim}{w}(\underset{\sim}{x},\underset{\sim}{t})\| \le W$ in S, and $\|\cdot\|_H$ is any suitable norm in R^n. If the $n \times n$ minors of A are non-singular we say that A satisfies the Haar condition.

Finally, we note that the algorithm is in fact applicable to the minimisation of $F(\underset{\sim}{f})$, where F is a general convex function bounded below. This case has application in the study of robust estimation.

2. Characterisation of stationary points

In this section, we develop representation results for the norm, and use these to characterise stationary values of $\|\underset{\sim}{f}\|$. An important role is played by the <u>dual norm</u>, defined by

$$\|\underset{\sim}{v}\|^* = \max_{\|\underset{\sim}{u}\| \le 1} \underset{\sim}{u}^T\underset{\sim}{v} \,. \tag{2.1}$$

Here, a continuous function is to be maximised on a compact set, so that there is a nonempty set of vectors at which the maximum is obtained. We say that these vectors are <u>aligned</u> with $\underset{\sim}{v}$. When the dual norm is prescribed, $\|\underset{\sim}{u}\|$ can be defined in an analogous fashion, and it is convenient to denote the set of vectors aligned with $\underset{\sim}{u}$ by $V(\underset{\sim}{u})$, so that

$$V(\underset{\sim}{u}) = \{\underset{\sim}{v} : \|\underset{\sim}{u}\| = \underset{\sim}{v}^T\underset{\sim}{u} , \quad \|\underset{\sim}{v}\|^* \leq 1\} . \qquad (2.2)$$

The set $V(\underset{\sim}{u})$ is also called the <u>subdifferential</u> of $\|\underset{\sim}{u}\|$ (Rockafeller (1970)). Geometrically, the aligned $\underset{\sim}{v}$ are normals to the supporting hyperplanes to $\|\cdot\|$ at $\underset{\sim}{u}$, and for this reason they are called subgradients. Note that $V(\underset{\sim}{u})$ is closed and convex.

<u>Examples</u> (i) If the norm is differentiable, then $V(\underset{\sim}{u})$ consists of the single vector $\nabla_{\underset{\sim}{u}}\|\underset{\sim}{u}\|$. This is readily seen by differentiating the relation $\|\alpha\underset{\sim}{u}\| = \alpha\|\underset{\sim}{u}\|$ valid for $\alpha \geq 0$ and then setting $\alpha = 1$. In particular, for the L_p norms $1 < p < \infty$,

$$v_i = u_i|u_i|^{p-2} \|\underset{\sim}{u}\|^{1-p} , \qquad i = 1,2,\ldots,m . \qquad (2.3)$$

(ii) For the L_∞ or maximum norm

$$V(\underset{\sim}{u}) = \text{conv}\{\text{sgn}(u_j)\underset{\sim}{e}_j , \forall u_j \ni |u_j| = \|\underset{\sim}{u}\| \} \qquad (2.4)$$

where conv denotes convex hull, and $\underset{\sim}{e}_j$ denotes the jth co-ordinate vector.

(iii) For the L_1 norm

$$V(\underset{\sim}{u}) = \{\underset{\sim}{v} : v_i = \text{sgn}(u_i) , \quad u_i \neq 0 \quad |v_i| \leq 1 , \quad u_i = 0\}.$$

We are now able to give necessary conditions for $\underset{\sim}{x}$ to be a local minimum of $\|\underset{\sim}{f}\|$. Two preliminary lemmas are stated without proof.

<u>Lemma 2.1</u> (Rockafellar (1970)). Let $\{\underset{\sim}{x}_i\}$ be a sequence of points tending to $\underset{\sim}{x}^*$. Then the limit points of $\{V(\underset{\sim}{f}(\underset{\sim}{x}_i))\}$ are contained in $V(\underset{\sim}{f}(\underset{\sim}{x}^*))$.

<u>Lemma 2.2</u> (Cheney (1966)). Let R be a closed, convex set. Then $0 \notin R$ iff $\exists \underset{\sim}{y} \ni \underset{\sim}{y}^T\underset{\sim}{x} < 0 , \forall \underset{\sim}{x} \in R$.

<u>Theorem 2.1</u> If $\underset{\sim}{x}$ is a local minimum of $\|\underset{\sim}{f}\|$, then $\exists \underset{\sim}{v} \in V(\underset{\sim}{f}(\underset{\sim}{x})) \ni \underset{\sim}{v}^TA = 0$.

<u>Proof</u>:- Let $\underset{\sim}{x}$ be a local minimum of $\|\underset{\sim}{f}\|$, let $R = \{\underset{\sim}{y} : \underset{\sim}{y} = A^T\underset{\sim}{v}, \underset{\sim}{v} \in V(\underset{\sim}{f}(\underset{\sim}{x}))\}$, and assume that $0 \notin R$. Then by lemma 2.2, $\exists \underset{\sim}{t} , \delta > 0 \ni \underset{\sim}{y}^T\underset{\sim}{t} < -\delta, \underset{\sim}{y} \in R$. We have

$$\begin{aligned}\|\underset{\sim}{f}(\underset{\sim}{x}+\lambda\underset{\sim}{t})\| &= \underset{\sim}{v}(\lambda)^T\underset{\sim}{f}(\underset{\sim}{x}+\lambda\underset{\sim}{t}) , \quad \underset{\sim}{v}(\lambda) \in V(\underset{\sim}{f}(\underset{\sim}{x}+\lambda\underset{\sim}{t})) \\ &= \underset{\sim}{v}(\lambda)^T\underset{\sim}{f}(x) + \lambda\underset{\sim}{v}(\lambda)^TA\underset{\sim}{t} + 0(\lambda^2) \\ &< \|\underset{\sim}{f}(\underset{\sim}{x})\| + \lambda(\underset{\sim}{v}(\lambda)^TA - \underset{\sim}{y}^T)\underset{\sim}{t} - \lambda\delta + 0(\lambda^2)\end{aligned}$$

for all $\underset{\sim}{y} \in R$.

By specialising to a sequence of values λ_i we can arrange that the sequence $\{A^T\underset{\sim}{v}(\lambda_i)\}$ converges to $\underset{\sim}{y}^* \in R$, using Lemma 2.1, and we have a contradiction for λ_i small enough.

Definition 2.1 $\underset{\sim}{x}$ is a stationary point of $\|\underset{\sim}{f}\|$ if $\exists\ \underset{\sim}{v} \in V(\underset{\sim}{f}(\underset{\sim}{x})) \ni \underset{\sim}{v}^T A = 0$.

Remark This generalises, in a natural fashion, the familiar definition of a stationary point in the calculus.

Definition 2.2 The LSP at $\underset{\sim}{x}$ is given by

$$\text{find } \underset{\sim}{h} \in R^n \text{ to minimise } \|\underset{\sim}{r}\| ,$$

$$\text{where } \underset{\sim}{r} = \underset{\sim}{f} + A\underset{\sim}{h} . \tag{2.5}$$

The LSP is a convex programming problem subject to linear constraints. It specialises to a linear least squares problem in L_2 , and to a linear programming problem in the case of polyhedral norms (Anderson and Osborne (1976)). Efficient methods for solving special cases have been known for some time. For the L_∞ problem Kelley (1958) and Stiefel (1959) considered an exchange algorithm, which was extended by Watson (1974) to include linear constraints. Recent developments include a descent algorithm due to Cline (1976). For the L_1 case, efficient algorithms have been given by Barrodale and Roberts (1973) and by Bartels et al (1977). In the L_p case, $p > 2$, Fletcher et al (1971) give an algorithm based on Newton's method, which reduces in the case $p = 2$ to solving the normal equations.

A consequence of Theorem 2.1 is that the solution of the LSP at $\underset{\sim}{x}$ is characterised by:

$$\exists\ \underset{\sim}{v} \in V(\underset{\sim}{r}) \ni \underset{\sim}{v}^T A = 0 . \tag{2.6}$$

Using convexity, it is easily seen that this condition is both necessary and sufficient for a global minimum of $\|\underset{\sim}{r}\|$.

Examples (i) If the norm is differentiable, the minimum occurs where

$$\nabla_{\underset{\sim}{h}} \|\underset{\sim}{r}\| = \nabla_{\underset{\sim}{r}} \|\underset{\sim}{r}\| A = \underset{\sim}{v}^T A = 0 .$$

(ii) For the L_∞ problem, it follows from the properties of linear programming that $\exists$ (n+1) indices $j \ni |r_j| = \|\underset{\sim}{r}\|$. It is conventional to call this set a reference, and we have

$$\underset{\sim}{v} = \sum_{j \in \text{ref}} \alpha_j \underset{\sim}{e}_j , \quad \sum_{j \in \text{ref}} |\alpha_j| = 1$$

and

$$\sum_{j \in \text{ref}} \alpha_j \rho_j(A) = 0$$

where $\rho_j(\cdot)$ denotes the jth row. Note that the Haar condition on A ensures that $\underset{\sim}{v}$ has at least (n+1) non-zero components.

(iii) Linear programming theory also gives information about the L_1 case. Here the condition that the optimum occurs at a vertex means that there must be (n+1) linearly independent vectors in $V(\underset{\sim}{r})$. This is only possible if there are n indices $j \ni r_j = 0$.

Note In the subsequent presentation references to $\underset{\sim}{r}$ will assume that it minimises the LSP. Clearly $\|\underset{\sim}{r}\| \leq \|\underset{\sim}{f}\|$.

Theorem 2.2 $\underset{\sim}{x}$ is a stationary point of $\|\underset{\sim}{f}\|$ iff

$$\|\underset{\sim}{r}\| = \|\underset{\sim}{f}\| \tag{2.7}$$

at $\underset{\sim}{x}$.

Proof:- Let (2.7) hold. Then, by (2.6) $\exists \ \underset{\sim}{v} \in V(\underset{\sim}{r}) \ni$

$$\underset{\sim}{v}^T \underset{\sim}{f} = \|\underset{\sim}{r}\| = \|\underset{\sim}{f}\|$$

$$\therefore \quad \underset{\sim}{v} \in V(\underset{\sim}{f})$$

so that $\underset{\sim}{x}$ is a stationary point. On the other hand assume $\underset{\sim}{x}$ is a stationary point. Then $\exists \ \underset{\sim}{v} \in V(\underset{\sim}{f}) \ni \underset{\sim}{v}^T A = 0$.

$$\therefore \quad \underset{\sim}{v}^T \underset{\sim}{f} - \|\underset{\sim}{f}\| - \underset{\sim}{v}^T \underset{\sim}{r} \leq \|\underset{\sim}{r}\| .$$

Remark This result reduces the characterisation property of a stationary point to a single scalar relation. An important feature of the algorithm described in the next section is that it sets out deliberately to produce a point satisfying this condition.

3. The generalised Gauss-Newton algorithm Let $\underset{\sim}{x}_i$ denote the current point reached by the algorithm, $\underset{\sim}{f}_i = \underset{\sim}{f}(\underset{\sim}{x}_i)$, $A_i = \nabla \underset{\sim}{f}(\underset{\sim}{x}_i)$. Then the basic form of the algorithm is

(i) Compute $\underset{\sim}{h}_i$, $\underset{\sim}{r}_i$ from LSP at $\underset{\sim}{x}_i$.

(ii) Choose γ_i as the largest member of the set $\{1, \theta, \theta^2, \ldots\}$ $0 < \theta < 1 \ni$

$$\psi(\underset{\sim}{x}_i, \gamma) = \frac{\|\underset{\sim}{f}_i\| - \|\underset{\sim}{f}(\underset{\sim}{x}_i + \gamma \underset{\sim}{h}_i)\|}{\gamma(\|\underset{\sim}{f}_i\| - \|\underset{\sim}{r}_i\|)} \geq \sigma > 0 . \tag{3.1}$$

(iii) $\underset{\sim}{x}_{i+1} = \underset{\sim}{x}_i + \gamma_i \underset{\sim}{h}_i$, $i := i+1$, repeat (i).

Note (i) If A has full rank in S then $\exists \ K \ni \|\underset{\sim}{h}\| \leq K \|\underset{\sim}{f}\| \ \forall \ \underset{\sim}{x} \in S$. In the subsequent analysis the boundedness of $\underset{\sim}{h}$ is assumed.

(ii) The particular form of line search is only one among a number of satisfactory strategies. The essential point is that γ_i must be chosen relatively large among the set of possible values.

(iii) If (3.1) is satisfied then the decrease in $\|\underset{\sim}{f}\|$ in the descent step is commensurate with the decrease obtained in the LSP. The parameter σ provides the basis for this comparison and is usually chosen small - say $\sigma = 10^{-4}$.

(iv) $\psi(\underset{\sim}{x},\gamma)$ is defined if $\underset{\sim}{x}$ is not a stationary point of $\|\underset{\sim}{f}\|$.

(v) If (3.1) is satisfied the sequence $\{\|\underset{\sim}{f}_i\|\}$ is decreasing and bounded below. It is therefore convergent.

We now prove the two key results that provide a basis for the utility of the algorithm.

Theorem 3.1 If $\|\underset{\sim}{r}_i\| < \|\underset{\sim}{f}_i\|$ $\exists\ \gamma > 0 \ni \psi(\underset{\sim}{x}_i,\gamma) \geq \sigma$.

Proof:- Using (1.2) and the form of the LSP we have

$$\underset{\sim}{f}(\underset{\sim}{x}_i+\gamma\underset{\sim}{h}_i) = \underset{\sim}{f}_i + \gamma A\underset{\sim}{h}_i + \gamma^2\|\underset{\sim}{h}_i\|^2\underset{\sim}{w}_i$$

$$= \underset{\sim}{f}_i + \gamma(\underset{\sim}{r}_i-\underset{\sim}{f}_i) + \gamma^2\|\underset{\sim}{h}_i\|^2\underset{\sim}{w}_i \ .$$

Thus, provided $0 \leq \gamma \leq 1$,

$$\|\underset{\sim}{f}(\underset{\sim}{x}_i+\gamma\underset{\sim}{h}_i)\| \leq (1-\gamma)\|\underset{\sim}{f}_i\| + \gamma\|\underset{\sim}{r}_i\| + \gamma^2\|\underset{\sim}{h}_i\|^2 W \ .$$

Thus

$$\psi(\underset{\sim}{x}_i,\gamma) \geq \frac{\gamma(\|\underset{\sim}{f}_i\| - \|\underset{\sim}{r}_i\|) - \gamma^2\|\underset{\sim}{h}_i\|^2 W}{\gamma(\|\underset{\sim}{f}_i\| - \|\underset{\sim}{r}_i\|)}$$

$$= 1 - \frac{\gamma\|\underset{\sim}{h}_i\|^2 W}{\|f_i\| - \|r_i\|} \qquad (3.2)$$

$$\to 1, \text{ as } \gamma \to 0 \text{ for fixed } i \ .$$

Remark This theorem shows that for each $\underset{\sim}{x}_i$, $\underset{\sim}{x}_i$ not a stationary point, we can find γ_i satisfying the test (3.1). It is important to note, however, that we cannot guarantee that the sequence $\{\gamma_i\}$ is bounded away from zero.

Theorem 3.2 $\|\underset{\sim}{f}_i\| - \|\underset{\sim}{r}_i\| \to 0$, $i \to \infty$.

Proof:- It is convenient to break the proof into two parts corresponding to the possibilities $\gamma_i \geq \hat{\gamma} > 0$, $\forall$ i and $\inf \gamma_i = 0$.

(i) If $\gamma_i \geq \hat{\gamma} > 0$ then from $\psi(\underset{\sim}{x}_i, \gamma_i) \geq \sigma$ we obtain

$$\frac{\|\underset{\sim}{f}_i\| - \|\underset{\sim}{f}_{i+1}\|}{\|\underset{\sim}{f}_i\| - \|\underset{\sim}{r}_i\|} \geq \hat{\gamma}\sigma \ .$$

$$\therefore \quad \|\underset{\sim}{f}_i\| - \|\underset{\sim}{r}_i\| \leq \frac{1}{\hat{\gamma}\sigma}(\|\underset{\sim}{f}_i\| - \|\underset{\sim}{f}_{i+1}\|)$$

$$\to 0 \ , i \to \infty$$

as the sequence $\{\|\underset{\sim}{f}_i\|\}$ converges.

(ii) If $\inf \gamma_i = 0$, then $\exists$ sequence $\{\gamma_i^0\} \to 0$ with the property that

$$\psi(\underset{\sim}{x}_i, \gamma_i^0) < \sigma$$

(for example, $\{\gamma_i^0\}$ can be chosen as a subsequence of $\{\gamma_i/\theta , \gamma_i < 1\})$. From (3.2) we obtain, for this subsequence,

$$1 - \frac{\gamma_i^0 \| \underset{\sim}{h}_i \|^2 W}{\| \underset{\sim}{f}_i \| - \| \underset{\sim}{r}_i \|} < \sigma \quad .$$

Thus

$$\| \underset{\sim}{f}_i \| - \| \underset{\sim}{r}_i \| < \frac{\gamma_i 0}{1-\sigma} W \| \underset{\sim}{h}_i \|^2$$

and the right hand side tends to zero with γ_i^0 as $\| \underset{\sim}{h}_i \|^2$ is assumed bounded.

<u>Corollary</u> The limit points of the sequence $\{\underset{\sim}{x}_i\}$ are stationary points of $\| \underset{\sim}{f} \|$.

4. Convergence of the full step method

Theorem 3.2 shows that the algorithm described in the previous section has a basic convergence property. However, the utility of the method depends on the rate of convergence being sufficiently rapid, and in particular, it is important that the step lengths be bounded away from zero. Therefore, it is of interest to determine conditions under which this occurs, and, since the approach is based on Newton's method it is natural to seek conditions which give rise to convergence of the method with $\gamma = 1$ (the full step method). This requires $\psi(\underset{\sim}{x}_i, 1) \geq \sigma$ for i sufficiently large, and, using equation (3.2), this will be satisfied provided that

$$\| \underset{\sim}{h}_i \|^2 \leq K(\| \underset{\sim}{f}_i \| - \| \underset{\sim}{r}_i \|) \ , \tag{4.1}$$

where K is sufficiently small.

A necessary condition for (4.1) is that $\| \underset{\sim}{h}_i \| \to 0$, $i \to \infty$, and this need not be satisfied, even if A_i has full rank (see Madsen (1975) for an example in the L_∞ norm). In fact the convergence of $\| \underset{\sim}{h}_i \|$ to zero requires that the LSP's have unique solutions which depend continuously on $\underset{\sim}{x}$. While necessary, this in itself is not in general sufficient, and we now turn to an analysis of conditions under which (4.1) can hold. These conditions, and the precise form of the resulting inequalities, lead naturally to consideration of two distinct classes of norms, which are conveniently treated separately in the next two sections.

5. Polyhedral norms

Consider the consistent set of linear inequalities

$$B\underset{\sim}{u} \leq \underset{\sim}{e} \tag{5.1}$$

where $\underset{\sim}{u} \in R^m$, B is an $N \times m$ matrix and $\underset{\sim}{e} \in R^N$ is a vector each component of which is unity. Then if

(i) $C = \{\underset{\sim}{u} : B\underset{\sim}{u} \le \underset{\sim}{e}\}$ is bounded and has a nonvoid interior,

(ii) $\underset{\sim}{u} \in C$ iff $-\underset{\sim}{u} \in C$,

the polyhedral norm of $\underset{\sim}{u}$ specified by C is defined by

$$\|\underset{\sim}{u}\| = \min\{\lambda : B\underset{\sim}{u} \in \lambda\underset{\sim}{e}\}. \tag{5.2}$$

Examples (i) The L_∞ norm in R^m is defined by choosing $B = \begin{bmatrix} I \\ -I \end{bmatrix}$.

(ii) The L_1 norm in R^m is defined by choosing B as the $2^m \times m$ matrix whose rows correspond to the 2^m different ways of filling m locations with either $+1$ or -1.

The Gauss-Newton method for polyhedral norms has been considered in detail by Anderson and Osborne (1977a), generalising the results of Osborne and Watson (1969, 1971). The fact that the polyhedral norm LSP is equivalent to a linear programming problem enables a convenient characterisation of the minimum to be given. Specifically, there exists a matrix B_i^σ consisting of $(n+1)$ of the rows of B such that $\rho_j(B_i^\sigma)\underset{\sim}{r}_i = \|\underset{\sim}{r}_i\|$, $j = 1,2,\ldots,n+1$, and a non-negative vector $\underset{\sim}{\mu}_i \in R^{n+1}$ such that

$$\underset{\sim}{\mu}_i^T B_i^\sigma A_i = 0$$

$$\underset{\sim}{\mu}_i^T \underset{\sim}{e} = 1 .$$

The multipliers $\underset{\sim}{\mu}_i$ play an important role in the derivation of the required convergence results, being required to satisfy a strong non-degeneracy assumption.

Definition 5.1 The multiplier condition is said to be satisfied on the compact set $T \subset S$ if

(i) the submatrices B_i^σ can be chosen to correspond to a fixed set of $(n+1)$ rows of B for all $\underset{\sim}{x}_i \in T$

$$\text{(ii)} \quad (\underset{\sim}{\mu}_i)_j \ge \hat{\mu} > 0 \quad \forall\, j,\ \underset{\sim}{x}_i \in T . \tag{5.3}$$

Theorem 5.1 (Anderson and Osborne (1977a))

If the multiplier condition is satisfied in T, then

(i) there exists a constant K such that

$$\|\underset{\sim}{h}_i\| \le \frac{K}{\hat{\mu}}(\|\underset{\sim}{f}_i\| - \|\underset{\sim}{r}_i\|) \tag{5.4}$$

for $\underset{\sim}{x}_i \in T$,

(ii) the ultimate rate of convergence of the full step method is at least of order 2.

Remark It is a direct consequence of (5.4) that the full step method is ultimately convergent.

A necessary condition for (5.3) is that the matrices $B_i^{\sigma}A_i$ satisfy the Haar condition in T . For the L_∞ norm, this reduces to the condition that the matrices A_i satisfy the Haar condition. It is, however, not possible to relax the conditions of the above theorem. In particular, an L_1 norm example in which the compactness condition is violated, with a resulting reduction in the rate of convergence to first order, is given in Anderson and Osborne (1977a).

6. Smooth, strictly convex, monotonic norms (SSM norms)

We now consider another family of norms, for each of whose members the modified Gauss-Newton algorithm exhibits similar performance characteristics. The norms of this class are typified by the L_p norms, $1 < p < \infty$, and the properties of the algorithm in this case are quite different from those for the polyhedral norms considered in the previous section. It turns out to be convenient to make use of twice differentiability of the norm where possible, and thus the L_p norms with $1 < p < 2$ are treated separately. In addition, we will see that the case $p = 2$ is a rather special one.

We begin by making precise the special properties possessed by SSM norms.

Definition 6.1 A norm $\|\cdot\|$ is smooth if $\exists$ a unique supporting hyperplane at each boundary point of the unit ball $\|\underset{\sim}{u}\| \leq 1$.

Definition 6.2 A norm $\|\cdot\|$ is strictly convex if

$$\|\underset{\sim}{u}\| + \|\underset{\sim}{v}\| = \|\underset{\sim}{u} + \underset{\sim}{v}\| \Rightarrow \underset{\sim}{u} = \lambda\underset{\sim}{v} .$$

Definition 6.3 A norm $\|\cdot\|$ is monotonic if

$$|u_i| \leq |v_i| , \; i = 1,2,\ldots,m \Rightarrow \|\underset{\sim}{u}\| \leq \|\underset{\sim}{v}\| .$$

A consequence of definition 6.1 is that smooth norms are differentiable, so that

$$V(\underset{\sim}{f}) = \{\underset{\sim}{v}\} = \{\nabla_{\underset{\sim}{f}}\|\underset{\sim}{f}\|\} .$$

In addition the components of $\underset{\sim}{v}$ for SSM norms are related to the components of $\underset{\sim}{f}$ in a rather simple way.

Lemma 6.1 Let $\underset{\sim}{v} \in V(\underset{\sim}{r})$. Then $\exists$ a diagonal matrix $U = \mathrm{diag}\{u_1,u_2,\ldots,u_m\} \ni$

$$\underset{\sim}{v} = U\underset{\sim}{r} \qquad (6.1)$$

with $u_i > 0$, $|r_i| > 0$, $u_i = 0$, $r_i = 0$.

This result is an elementary consequence of the properties of SSM norms and the proof is omitted.

Example L_p norms $1 < p < \infty$

$$u_i = |r_i|^{p-2} \, \|\underset{\sim}{r}\|^{1-p} \quad , \quad r_i \neq 0$$

$$= 0 \qquad\qquad , \quad r_i = 0 \; .$$

An immediate consequence of Lemma 6.1 is that the solution to the linear subproblem is characterised by

$$A^T U \underset{\sim}{r} = 0$$

so that

$$A^T U A \underset{\sim}{h} = - A^T U \underset{\sim}{f} \; . \tag{6.2}$$

This equation generalises the normal equations which hold if $p = 2$ (the case $U = \frac{1}{\|\underset{\sim}{r}\|} I$). However, if $p \neq 2$ then a stronger condition than A full rank is required to guarantee that the matrix $A^T U A$ is nonsingular. A sufficient condition is that A satisfies the Haar condition, for then U must have at least $(n+1)$ non-zero components if $\|\underset{\sim}{r}\| \neq 0$.

Definition 6.4 A SSM norm satisfies condition A if $\exists$ a smooth norm $\|\cdot\|_A$ such that

$$\|\underset{\sim}{w}(\underset{\sim}{f})\|_A = \|\underset{\sim}{f}\|^2 \tag{6.3}$$

where $\qquad \underset{\sim}{w}(\underset{\sim}{f})^T = \{f_1^{\,2}, f_2^{\,2}, \ldots, f_m^{\,2}\}$.

Example The L_p norms satisfy condition A for $p > 2$. In this case $\|\cdot\|_A = \|\cdot\|_{p/2}$.

Lemma 6.2 Let $\|\underset{\sim}{w}(\underset{\sim}{r})\|_A = \underset{\sim}{g}^T \underset{\sim}{w}(\underset{\sim}{r})$, $\|\underset{\sim}{g}\|_A^* \leq 1$. Then

$$\underset{\sim}{g} = \|\underset{\sim}{r}\| \underset{\sim}{u}$$

where $\qquad \underset{\sim}{v} = U \underset{\sim}{r}$.

Proof:- From

$$\|\underset{\sim}{w}(\underset{\sim}{r})\|_A = \|\underset{\sim}{r}\|^2$$

it follows on differentiating both sides with respect to $\underset{\sim}{r}$ that

$$\underset{\sim}{g}^T \begin{pmatrix} 2r_1 & & \\ & \ddots & \\ & & 2r_m \end{pmatrix} = 2 \|\underset{\sim}{r}\| \, \underset{\sim}{v}^T \; .$$

Corollary $\qquad \|\underset{\sim}{r}\| \; \|\underset{\sim}{u}\|_A^* \leq 1$.

Note For norms satisfying condition A the elements of U are given up to a scalar multiplier by the components of $\underset{\sim}{g}$ and thus correspond to the components of an aligned vector for a particular smooth norm.

In what follows it will be convenient to drop references to the subscript i of the point $\underset{\sim}{r}_i$ at which the LSP is defined.

Theorem 6.1 Let the norm satisfy condition A . If

$$\|\underset{\sim}{r}\| \left(\min_{\|\underset{\sim}{t}\| = 1} \underset{\sim}{t}^T A^T U A \underset{\sim}{t} \right) > \Delta > 0 \tag{6.4}$$

then

$$\|\underset{\sim}{h}\|^2 \leq \frac{2}{\Delta} \|\underset{\sim}{f}\| (\|\underset{\sim}{f}\| - \|\underset{\sim}{r}\|) . \tag{6.5}$$

Proof:- From equation (6.2) and the statement of the LSP we have

$$\underset{\sim}{h}^T A^T U A \underset{\sim}{h} = \underset{\sim}{f}^T U \underset{\sim}{f} - \|\underset{\sim}{r}\| .$$

Now

$$\underset{\sim}{f}^T U \underset{\sim}{f} = \sum_{i=1}^{m} f_i^2 u_i$$

$$\leq \|\underset{\sim}{f}\|^2 \|\underset{\sim}{u}\|_A{}^*$$

$$\leq \|\underset{\sim}{f}\|^2 / \|\underset{\sim}{r}\|$$

by the Corollary to Lemma 6.2. Thus

$$\underset{\sim}{h}^T A^T U A \underset{\sim}{h} \leq \frac{1}{\|\underset{\sim}{r}\|} (\|\underset{\sim}{f}\|^2 - \|\underset{\sim}{r}\|^2)$$

$$\leq \frac{2\|\underset{\sim}{f}\|}{\|\underset{\sim}{r}\|} (\|\underset{\sim}{f}\| - \|\underset{\sim}{r}\|)$$

and the desired result follows.

Remark This result holds also for the L_2 norm where we can make use of the special form of U . In this case we find we require $\Delta \leq \lambda_{min}(A^T A)$, the smallest eigenvalue of $A^T A$.

A bound for Δ can also be given for the case $p > 2$, and this is stated in the following theorem. The proof is elementary and is omitted.

Theorem 6.2 Assume $p > 2$, $\|\underset{\sim}{r}\| \neq 0$. Let $\{B_j\}$ be the set of $n \times n$ submatrices of A . If

(i) A satisfies the Haar condition,

(ii) $\min_j \min_{\|\underset{\sim}{t}\|=1} \underset{\sim}{t}^T B_j^T B_j \underset{\sim}{t} \geq \delta^2 > 0$, and

(iii) $\|\underset{\sim}{r}\| / |r_n| \leq K$, where the components of $\underset{\sim}{r}$ are ordered so that

$$|r_1| \geq |r_2| \geq \ldots \geq |r_n| \geq |r_j|, \quad j = n+1,\ldots,m,$$

then we can take

$$\Delta = K^{2-p} \delta^2 .$$

Remark Conditions (ii) and (iii) are consequences of the Haar condition at each point, although we need to strengthen the pointwise properties to obtain a uniform result. Note also that $\|\underset{\sim}{r}\| / |r_n| \to 1$ as $p \to \infty$.

For $1 < p \leq 2$, the L_p norms do not satisfy condition A . A similar inequality to (6.5) still holds, however, and this is the substance of the following theorem.

Theorem 6.3 Let A satisfy the Haar condition, let the components of $\underset{\sim}{r}$ be ordered so that

$$r_j f_j \geq 0 \quad , \quad r_j + f_j \neq 0 \quad , \quad j = 1,2,\ldots,k,$$

and let

$$\sum_{j=1}^{k} (\rho_j(A)\ \underset{\sim}{t})^2 \geq \delta^2 > 0$$

where $\underset{\sim}{h} = \|\underset{\sim}{h}\|\ \underset{\sim}{t}$. Then, if $\|\underset{\sim}{r}\| \neq 0$,

$$\|\underset{\sim}{h}\|^2 \leq \frac{2\|\underset{\sim}{f}\|}{(p-1)\delta^2} (\|\underset{\sim}{f}\| - \|\underset{\sim}{r}\|) . \tag{6.6}$$

Proof:- It is an easy consequence of the Haar condition that $k \geq 1$ and $\delta > 0$ at each point. For $j \in [1,k]$, $|\theta f_j + (1-\theta) r_j|^p$ is differentiable for $0 < \theta < 1$ and we can write

$$|f_j|^p = |r_j|^p - p r_j |r_j|^{p-2} (\rho_j(A)\ h)$$
$$+ \tfrac{1}{2} p(p-1)\ (\rho_j(A)\ h)^2 \quad |\bar{r}_j|^{p-2}$$

where the bar denotes a mean value in $0 < \theta < 1$. We can ignore indices such that $f_j = r_j = 0$, and for the remaining indices convexity gives

$$|f_j|^p \geq |r_j|^p - p r_j |r_j|^{p-2} (\rho_j(A) \underset{\sim}{h}) .$$

Now

$$|\bar{r}_j| = |\theta f_j + (1-\theta)r_j| \ , \ 0 < \theta < 1 \ , \ j \in [1,k]$$
$$\leq \theta \| \underset{\sim}{f} \| + (1-\theta) \| \underset{\sim}{r} \| \leq \| \underset{\sim}{f} \|$$

$$\therefore \quad |\bar{r}_j|^{p-2} \geq \| \underset{\sim}{f} \|^{p-2} .$$

Summing over all j we obtain

$$\| \underset{\sim}{f} \|^p \geq \| \underset{\sim}{r} \|^p - p \sum_{j=1}^{m} r_j |r_j|^{p-2} \rho_j(A) \underset{\sim}{h}$$
$$+ \tfrac{1}{2} p(p-1) \| \underset{\sim}{f} \|^{p-2} \sum_{j=1}^{k} (\rho_j(A) \underset{\sim}{h})^2$$
$$= \| \underset{\sim}{r} \|^p + \tfrac{1}{2} p(p-1) \| \underset{\sim}{f} \|^{p-2} \sum_{j=1}^{k} (\rho_j(A) \underset{\sim}{h})^2 ,$$

using the condition (2.6) for the minimum of the LSP. Thus

$$\| \underset{\sim}{h} \|^2 \leq \frac{2}{p(p-1)\delta^2} \| \underset{\sim}{f} \|^{2-p} (\| \underset{\sim}{f} \|^p - \| \underset{\sim}{r} \|^p)$$
$$\leq \frac{2 \ \| \underset{\sim}{f} \|}{(p-1)\delta^2} (\| \underset{\sim}{f} \| - \| \underset{\sim}{r} \|) .$$

Remark We have assumed that $\| \underset{\sim}{r} \| \neq 0$ in deriving the inequalities (6.5) and (6.6). However, inequalities of the required form follow readily when $\| \underset{\sim}{r} \| = 0$.

Provided that the inequalities (6.5) and (6.6) hold uniformly in a neighbourhood of the solution they are sufficient to guarantee the convergence of the full step method only if $\| \underset{\sim}{f} \|$ is sufficiently small. This is in marked contrast to the polyhedral norm case. This contrast persists also in the rate of convergence results for SSM norms.

Theorem 6.4 Let the matrix U defined in equation (6.1) have components that are smooth functions of $\underset{\sim}{x}$. Then the rate of convergence of the full step method is at best first order unless $\| \underset{\sim}{f} \| \to 0$.

Remark The condition on U is satisfied if $3 < p < \infty$, or $p = 2$.

Proof:- Let $\underset{\sim}{x}^*$ be a stationary point. Then

$$\underset{\sim}{f}(\underset{\sim}{x}^*) - \underset{\sim}{f}_i + A_i(\underset{\sim}{x}^* - \underset{\sim}{x}_i) + \| \underset{\sim}{x}^* - \underset{\sim}{x}_i \|^2 \underset{\sim}{w}_i$$

$$\therefore \quad \| \underset{\sim}{f}_i + A_i(\underset{\sim}{x}^* - \underset{\sim}{x}_i) \| \leq \| \underset{\sim}{f}(\underset{\sim}{x}^*) \| + W \| \underset{\sim}{x}^* - \underset{\sim}{x}_i \|^2$$

$$\therefore \quad \| \underset{\sim}{r}_i \| \leq \| \underset{\sim}{f}(\underset{\sim}{x}^*) \| + W \| \underset{\sim}{x}^* - \underset{\sim}{x}_i \|^2 . \tag{6.7}$$

Now
$$\underset{\sim}{f}_i + A_i \underset{\sim}{h}_i = \underset{\sim}{r}_i$$

so that
$$A_i(\underset{\sim}{x}^* - \underset{\sim}{x}_{i+1}) = \underset{\sim}{f}(\underset{\sim}{x}^*) - \underset{\sim}{r}_i - \| \underset{\sim}{x}^* - \underset{\sim}{x}_i \|^2 \underset{\sim}{w}_i . \tag{6.8}$$

Thus

$$A_i^T U_i A_i (\underset{\sim}{x}^* - \underset{\sim}{x}_{i+1}) = A_i^T U_i \underset{\sim}{f}(\underset{\sim}{x}^*) - \|\underset{\sim}{x}^* - \underset{\sim}{x}_i\|^2 A_i^T U_i \underset{\sim}{w}_i$$

$$= \{A_i^T U_i - A^{*T} U^*\} \underset{\sim}{f}(\underset{\sim}{x}^*) - \|\underset{\sim}{x}^* - \underset{\sim}{x}_i\|^2 A_i^T U_i \underset{\sim}{w}_i$$

where the * denotes evaluation at $\underset{\sim}{x}^*$, and where we have used

$$A^{*T} U^* \underset{\sim}{f}(\underset{\sim}{x}^*) = A^{*T} U^* \underset{\sim}{r}^* = 0 .$$

It follows that

$$A_i^T U_i A_i (\underset{\sim}{x}^* - \underset{\sim}{x}_{i+1}) = \|\underset{\sim}{x}^* - \underset{\sim}{x}_i\| \frac{d}{d\underset{\sim}{t}} \overline{(A^T U)} \underset{\sim}{f}(\underset{\sim}{x}^*) - \|\underset{\sim}{x}^* - \underset{\sim}{x}_i\|^2 A_i^T U_i \underset{\sim}{w}_i$$

where the bar denotes a mean value, and $\underset{\sim}{t}$ is a unit vector in the direction $(\underset{\sim}{x}_i - \underset{\sim}{x}^*)$.

<u>Remark</u> If $\underset{\sim}{f}(\underset{\sim}{x}^*) = 0$ then second order convergence follows directly from equations (6.7) and (6.8) and is thus valid for all norms. In this case it only requires A to have full rank.

7. <u>Modifying the LSP</u>

As shown in Section 3, an extremely important requirement for convergence of the Gauss-Newton method is that the matrices A_i have full rank, and this is a condition which may well be violated in practice. In order to achieve robustness, alternative LSP's have been proposed: for example, for the least squares problem search directions $\underset{\sim}{h}$ can be chosen by minimising the L_2 norm of the vector

$$\underset{\sim}{r} = \begin{bmatrix} \underset{\sim}{r}_1 \\ \underset{\sim}{r}_2 \end{bmatrix} = \begin{bmatrix} \underset{\sim}{f} \\ \underset{\sim}{0} \end{bmatrix} + \begin{bmatrix} A \\ B \end{bmatrix} \underset{\sim}{h} ,$$

where B is an $n \times n$ matrix chosen so that $\begin{bmatrix} A \\ B \end{bmatrix}$ has full rank. The particular choice $B = \lambda I$ gives the Levenberg method, and here the value of λ can be used to control the size of $\|\underset{\sim}{h}\|_H$ so that an auxiliary line search parameter is not required. This approach has been shown to be capable of yielding algorithms with excellent global convergence properties (Osborne (1977)).

The application of a similar approach for the L_∞ norm has been considered by Madsen (1975), and for general polyhedral norms by Anderson and Osborne (1977b). These algorithms obtain search directions by minimising the particular norm of the modified residual defined by

$$\|\underset{\sim}{r}\|_A = \max\{\|\underset{\sim}{r}_1\| , \lambda\|\underset{\sim}{h}\|_H\} .$$

Such methods appear to be extremely valuable, and we conclude by stating a general result indicating a wider class of possible choices of norm which can be used for the modified residual vector.

Theorem 7.1 Let $\underset{\sim}{v} \in V(\underset{\sim}{r})$ be partitioned in an obvious way so that

$$\| \underset{\sim}{r} \|_A = \underset{\sim}{v}_1^T \underset{\sim}{r}_1 + \underset{\sim}{v}_2^T \underset{\sim}{r}_2 .$$

Then convergence to a stationary point of $\| \underset{\sim}{f} \|$ can be obtained provided that $\| \underset{\sim}{r} \|_A$ is such that

$$\| \underset{\sim}{v}_2 \|_H \to 0 \quad \text{as} \quad \| \underset{\sim}{r}_2 \|_H \to 0 .$$

Remark This may readily be seen to be true for all SSM norms, and for the L_∞ norm, but it is not true for the L_1 norm.

References

Anderson, D H and M R Osborne (1976). Discrete, linear approximation problems in polyhedral norms, Num. Math. 26, 179-189.

Anderson, D H and M R Osborne (1977a). Discrete, non-linear approximation problems in polyhedral norms, Num. Math. 28, 143-156.

Anderson, D H and M R Osborne (1977b). Discrete, non-linear approximation problems in polyhedral norms : a Levenberg-like algorithm. Num. Math. 28, 157-170.

Barrodale, I and F D K Roberts (1973). An improved algorithm for discrete ℓ_1 linear approximation, SIAMJ Num. Anal. 10, 839-848.

Bartels, R H, A R Conn and J W Sinclair (1977). Minimisation techniques for piecewise differentiable functions - the ℓ_1 solution to an over-determined linear system, SIAMJ Num. Anal. (to appear).

Cheney, E W (1966). Introduction to Approximation Theory, McGraw-Hill, New York.

Cline, A K (1976). A descent method for the uniform solution to overdetermined systems of linear equations, SIAMJ Num. Anal. 13, 293-309.

Fletcher, R, J A Grant and M D Hebden (1971). The calculation of linear best L_p approximations, Computer J. 14, 276-279.

Kelley, J E (Jr) (1958). An application of linear programming to curve fitting, SIAMJ 6, 15-22.

Madsen, K (1975). An algorithm for minimax solutions of overdetermined systems of non-linear equations, JIMA 16, 321-328.

Osborne, M R (1972). An algorithm for discrete, non-linear best approximation problems: In: Numerische Methoden der Approximationstheorie, Band 1, eds. L Collatz and G Meinardus. Birkhauser-Verlag.

Osborne, M R (1977). Nonlinear least squares - the Levenberg algorithm revisited, J. Aust. Math. Soc., Series B (to appear).

Osborne, M R and G A Watson (1969). An algorithm for minimax approximation in the non-linear case, Computer J. 12, 64-69.

Osborne, M R and G A Watson (1971). On an algorithm for non-linear L_1 approximation, Computer J. 14, 184-188.

Rockafellar, R T (1970). Convex Analysis, Princeton, New Jersey. Princeton Univ. Press.

Stiefel, E L (1959). Über diskrete und lineare Tschebyscheff-Approximationen, Num. Math. 1, 1-28.

Watson, G A (1974). The calculation of best restricted approximations, SIAMJ Num. Anal. 11, 693-699.

FINITE DIFFERENCE SOLUTION OF TWO-POINT BOUNDARY VALUE PROBLEMS AND SYMBOLIC MANIPULATION

V. Pereyra*

1. Introduction

Great emphasis has been put in recent years on the approximate solution of two-point boundary value problems for first order systems of differential equations. A number of authors have pointed out the advantages of doing so: mainly its generality, algorithmic simplicity, and the natural way in which arbitrary nonuniform meshes can be dealt with (cf. Keller [1974], Lentini and Pereyra [1977]).

None of these arguments have lost their standing, but we will show in this paper that a great deal of efficiency, both computational and storage wise, can be gained by treating high order systems as they come, instead of transforming them into larger first order systems.

The price to be paid for this efficiency is higher complexity, and in order to generate the wealth of finite difference formulae necessary to carry out our project in some generality we have had to resort to the use of an algebraic manipulation system: MACSYMA [1975].

The purpose of this paper is then to exhibit explicit finite difference formulae for high order linear differential systems subject to two-point boundary conditions. The basic schemes will be consistent, as compact as possible in the sense of Kreiss [1972], centered, and therefore they will be stable and second order accurate. If the data is regular it then will exist asymptotic expansions for the global error and therefore Richardson extrapolations or deferred corrections can be used to increase the accuracy (cf. Keller [1974], Pereyra [1967], Kreiss [1972]).

* This work was supported under Contract No. AT-04-3-767, Project Agreement 12 with Energy Research and Development Administration.

In Section 4 we present an explicit deferred correction algorithm. Complete details and additional results can be found in Keller and Pereyra [1977a, 1977b].

2. THE PROBLEM AND ITS DISCRETIZATION

We consider the general n-th order linear systems

$$\mathcal{L}\, y(t) \equiv y^{(n)} + \sum_{\nu=0}^{n-1} A_\nu(t)\, y^{(\nu)}(t) = f(t) \quad , \quad a \leq t \leq b \quad , \tag{2.1}$$

where y, f are d-vector functions and the $A_\nu(t)$ are $d \times d$ matrix functions. We will assume that f and A_ν are $C^N(c,d)$ functions, for N sufficiently large and $(c,d) \supset [a,b]$, and that (2.1) is subject to the nd linear boundary conditions

$$B_k y \equiv \sum_{\nu=0}^{n-1} [\, B_{k\nu}(a)\, y^{(\nu)}(a) + B_{k\nu}(b)\, y^{(\nu)}(b)] = g_k \quad , \quad 0 \leq k \leq n-1 \quad , \tag{2.2}$$

where the $B_{k\nu}(a)$, $B_{k\nu}(b)$ are $d \times d$ matrices. We assume that this nd linear constraints are independent and that the boundary value problem (2.1)-(2.2) has a unique smooth solution in (c,d).

Centered Compact Difference Schemes

To obtain centered schemes we must treat system (2.1) according to the parity of its order. We will concentrate here on the description of the even order case $n = 2m$. A more detailed description of both even and odd order cases, together with proofs will appear elsewhere (Keller and Pereyra [1977a]).

On the interval (c,d) we introduce a uniform net defined by

$$\begin{aligned} t_j &= a + (j-\tfrac{1}{2})h \quad , \quad h = (b-a)/J \quad , \\ j &= -r\, , \; -r+1, \; \ldots, \; J+r \quad . \end{aligned} \tag{2.3}$$

The integer r will be fixed independently of h and therefore we will need to go outside of the interval $[a,b]$ only a few mesh points. As we will see below, this is done to allow the use of centered approximations to (2.1) over the whole interval (a,b), and also to approximate the boundary conditions (2.2) and to obtain the deferred correction operators. In this last respect, the present algorithm corrects a long standing and irritating defect of earlier implementations of deferred corrections for boundary value problems (see also Pereyra, Proskurowski and Widlund [1977] for a different cure to the same ailment). Actually the seeds of these ideas have been around for a long time as one can see in Fox [1957].

We introduce now some basic difference operators:

$$D_+ v_j \equiv h^{-1}(v_{j+1} - v_j) \;;\; D_- v_j \equiv h^{-1}(v_j - v_{j-1}) \tag{2.4}$$
$$D_0 v_j \equiv (2h)^{-1}(v_{j+1} - v_{j-1}) \;;\; M_+ v_j \equiv \tfrac{1}{2}(v_{j+1} + v_j) \; .$$

The discretizations we consider are made up with powers of these operators and for $n = 2m$ they have the form:

$$\mathcal{L}_h u_j \equiv (D_+D_-)^m u_j + \sum_{\nu=0}^{m-1} [A_{2\nu}(t_j)(D_+D_-)^\nu u_j \tag{2.5}$$
$$+ A_{2\nu+1}(t_j)(D_+D_-)^m D_0 u_j] = f(t_j) \; ,$$
$$1 \leqslant j \leqslant J \; ,$$

$$B_k^h u^h \equiv \sum_{\nu=0}^{m-1} \{[B_{k,\,2\nu+1}(a)(D_+D_-)^\nu D_+ u_0 + \tag{2.6}$$
$$B_{k,\,2\nu+1}(b)(D_+D_-)^\nu D_+ u_J] + [B_{k,\,2\nu}(a)(D_+D_-)^\nu M_+ u_0 +$$
$$B_{k,\,2\nu}(b)(D_+D_-)^\nu M_+ u_J]\} = g_k \; , \quad 0 \leqslant k \leqslant n-1 \; .$$

These are second order consistent, as compact as possible (i.e. they use at most $n+1$ mesh points) approximations and the general theory of Kreiss [1972] guarantees that they will be stable and therefore convergent of order h^2.

In terms of ordinates (2.5) and (2.6) will take the form

$$\mathcal{L}_h u_j \equiv \sum_{s=-m}^{m} C_s(t_j;h)u_{j+s} = f_j$$

$$B_k^h u^h \equiv \sum_{s=-(m-1)}^{m} [C_{ks}(a;h)u_s + C_{ks}(b;h)u_{J+s}] = g_k$$

where C_s, $C_{ks}(a;h)$, $C_{ks}(b;h)$ are $d \times d$ matrices.

We would like to make these matrix coefficients more explicit, so that they can be computed. Writing the difference operators in terms of ordinates we obtain

$$D_{2,e}^{2\nu} u_i \equiv (D_+D_-)^\nu u_i \equiv h^{-2\nu} \sum_{s=-\nu}^{\nu} \omega_s^{2\nu,e} u_{i+s}$$

$$D_{2,e}^{2\nu+1} u_i \equiv (D_+D_-)^\nu D_0 u_i \equiv h^{-(2\nu+1)} \sum_{s=-(\nu+1)}^{\nu+1} \omega_s^{2\nu+1,e} u_{i+s} .$$

Replacing in (2.5), and reordering we obtain

(2.7)
$$C_s(t_j;h) = \omega_s^{n,e} I + \sum_{\nu=|s|}^{m-1} h^{-2\nu} A_{2\nu}(t_j)\omega_s^{2\nu,e}$$
$$+ \sum_{\nu=\max(0,|s|-1)}^{m-1} h^{-(2\nu+1)} A_{2\nu+1}(t_j)\omega_s^{2\nu+1,e}$$

$$s = -m, \dots, m \; ; \; j = 1, \dots, J ,$$

and similarly for the boundary conditions.

Thus the problem reduces to knowing the weights $\omega_s^{\mu,e}$ for various values of μ. For small values of μ they are standard and can be found in many places. It is unlikely that any physical application will ever require μ to be larger than 4, so obtaining these weights does not present any new problem. With them we can compute the matrix coefficients in our difference approximations and assembling them together we will arrive at a linear system of equations that the unknown mesh function u^h must satisfy

(2.8)
$$\mathcal{A} u^h = f^h .$$

$\mathcal{A}$ is a band matrix which in block form has the following aspect

$$
(2.9) \qquad \mathcal{A} = \begin{bmatrix}
C_{1,-m+1} & C_{1,-m+2} & \cdots & C_{1m} & 0 & \cdots & 0 \\
\cdot & \cdot & \cdots & \cdot & \cdot & \cdots & \cdot \\
C_{p,-m+1} & C_{p,-m+2} & \cdots & C_{pm} & 0 & \cdots & \cdot \\
C_{1,-m} & C_{1,-m+1} & \cdots & C_{1,m-1} & C_{1,m} & 0 \;\cdot & 0 \\
0 & C_{2,-m} & \cdots & \cdot & \cdot & C_{2m} & 0 \\
0 & & & & & & \\
 & & C_{J,-m} & \cdot \quad \cdot & \cdot & & C_{J,m} \\
 & \cdot & 0 & & C_{p+1,-m+1} & \cdots & C_{p+1,m} \\
0 & & & & \cdot \;\cdot \;\cdot \;\cdot & & \cdot \;\cdot \;\cdot \\
 & & & & C_{nd,-m+1} & \cdots & C_{nd,m}
\end{bmatrix}
$$

where we have assumed separated boundary conditions, p of them initial conditions and $nd - p$ end conditions. The scalar half band width β is

$$
(2.10) \qquad \beta = \max(nd\ ,\ p+d\ ,\ (n+1)d-p) - 1\ .
$$

3. SOLUTION OF THE LINEAR EQUATIONS

At this stage most of the complexity is concentrated on solving the large, sparse system of linear equations (2.8). The data will have to be evaluated over the whole mesh $\{t_j\}$ whatever algorithm we use, so any possible savings will come from the algorithm used for these linear equations.

We have considered two possibilities for the high order systems:

a) A standard scalar band solver with partial pivoting.

b) A Gaussian elimination code with row pivoting for "almost block diagonal systems" by de Boor and Weiss [1976].

The number of arithmetic operations for (a) on a system with dJ equations is essentially

$$
(3.1) \qquad O_n^{(a)} \equiv 2n\ Jd\beta^2
$$

while for (b) we obtain in our case

$$O_n^{(b)} = \frac{Jd^2}{2}((2n+1)p+d) \approx Jd^2\ np \quad . \tag{3.2}$$

In terms of storage we have respectively

$$S_n^{(a)} = Jd(2\beta + 1) \quad , \tag{3.3}$$

$$S_n^{(b)} = J(n+1)d\ (p+d) \quad . \tag{3.4}$$

In order to be able to compare these quantities more easily we will assume now that $d \leqslant p \leqslant (n-1)d$, in which case $\beta = nd - 1$ and

$$\tilde{O}_n^{(a)} \equiv 2J\ n^2 d^3 \quad , \tag{3.5}$$

$$\tilde{S}_n^{(a)} = 2J\ nd^2 \quad . \tag{3.6}$$

Thus we see that the work and storage ratios are essentially

$$W^{b,a} \equiv \frac{O_n^{(b)}}{\tilde{O}_n^{(a)}} = \frac{p}{2nd} \quad , \tag{3.7}$$

$$R^{b,a} \equiv \frac{S_n^{(b)}}{\tilde{S}_n^{(a)}} = \left(\frac{p}{2d} + \frac{1}{2}\right)\frac{n+1}{n} \quad . \tag{3.8}$$

Method (b) will always require fewer arithmetic operations than method (a), but for $d \leqslant p$ it will take more storage. So our choice of algorithm will be guided by the type of computer installation constraints we may have.

Let us now consider the equivalent first order system with $n \times d$ equations associated with (2.1)-(2.2).

For first order systems we have available another linear equations solver: the alternate pivoting algorithm for block tridiagonal systems (B3D) described in Keller [1974] and used in Lentini and Pereyra [1975] (see also Pereyra and Lee [1977]). This algorithm is optimal in terms of space and its counts are

(3.9) $$O^{B3D} = Jd^3 n^3$$

(3.10) $$S^{B3D} = 2Jn^2 d^2 \quad .$$

The scalar band solver (method (a)) is not competitive for first order systems since it takes 3/2 as much storage and 4.5 as much work than the B3D method.

We give now some operation and storage ratios for the various methods.

The count for method (b) on a first order system with (nd) equations is obtained from (3.2), (3.4) by replacing n by 1 and d by nd:

(3.11) $$O_1^{(b)} = \frac{Jn^2 d^2}{2} (3p + nd) \quad ,$$

(3.12) $$S_1^{(b)} = 2\, J\, n\, d(p + nd) \quad .$$

Taking as a new parameter $r = p/nd$ we compare now method (b) with method B3D and we get the ratios

(3.13) $$W^{B3D,\, b} = \frac{2}{3r + 1} \quad ,$$

(3.14) $$R^{B3D,\, b} = \frac{1}{r + 1} \quad .$$

Since $0 < r < 1$, this says that method (b) always takes more storage than method B3D, and that it takes less arithmetic operations only if $\frac{2}{3r + 1} > 1$, i.e. if $r < 1/3$ or if $p < nd/3$.

Thus, for $p \geq nd/3$ method B3D is to be preferred, and for $p < nd/3$ one must balance the gain in computational speed against the loss in storage in order to choose the most appropriate method.

However, our main interest now is to compare the best method for an nth order system with that for its equivalent first order system. Thus we list now the work and storage ratios for method (b) as an nth order solver as compared to methods B3D and (b) again as a first order system solver:

$$W^{b,b}_{1,n} = \frac{O^{(b)}_1}{O^{(b)}_n} = \frac{n}{2r} + \frac{3n}{2} ,$$

(3.15)

$$R^{b,b}_{1,n} = \frac{S^{(b)}_1}{S^{(b)}_n} = 2 + 2/r$$

(3.16) $$W^{B3D,b}_{1,n} = \frac{n}{r} , \quad S^{B3D,b}_{1,n} = 2/r .$$

Formulas (3.15)-(3.16) show some impressive gains to be obtained if one solves the problem as an nth order system. For instance, for $n = 4$, $r = \frac{1}{2}$ we gain a factor 8 in speed by using method (b) as compared to B3D, while using only 1/4 as much storage.

For $n = 4$, $r = 1/4$ we use (b) as a first order solver since it is faster, but still (b) as an nth order solver will run 14 times faster and use 1/10 as much storage.

In conclusion, for first order systems the best solvers are Keller's Block Tridiagonal solver with alternate pivoting and de Boor and Weiss almost block diagonal solver.

For high order systems, it is best to solve them as they come using the methods of this paper instead of transforming them to first order. If storage is no concern then the de Boor and Weiss algorithm should be used since it is faster. However, if storage is a problem then the slower but less space consumming scalar band solver with partial pivoting should be preferred.

4. HIGHER ORDER METHODS

We would like to obtain now methods with orders of accuracy higher than 2, while preserving the structure of the linear systems (2.8) to be solved. An efficient way of doing this is by deferred corrections.

If we use the exact solution $y(t)$ in (2.5)-(2.6) we obtain the local truncation error:

(4.1) $$\tau_h \equiv \mathcal{L}_h[y]$$

and similarly for B^h_k. Expanding in Taylor's series we get

$$\tau_{hj} = \sum_{\nu=1}^{L} h^{2\nu}\, T_\nu[y_j] + \mathcal{O}(h^{2L+2}) , \tag{4.2}$$

where

$$T_\nu[y_j] = \sum_{\mu=0}^{n} \alpha^e_{\nu\mu}\, A_\mu(t_j)\, y^{(\mu+2\nu)}(t_j) . \tag{4.3}$$

An important consequence of using exclusively centered formulae is that the expansion (4.2) has only even powers of h.

The coefficients $\alpha^e_{\nu\mu}\ y_j^{(\mu+2\nu)}$ of (4.3) can be obtained from the expansions of the basic difference operators (2.4) with the aid of MACSYMA. For instance, for even $\mu = 2k$, we start from the well known expansion for (D_+D_-):

$$(D_+D_-)y_j \equiv D^2_{2,e}\, y_j = [\, D^2 + \sum_{\nu=1}^{L} \frac{2D^{2\nu+2}}{(2\nu+2)!}\, h^{2\nu}]\, y_j + \mathcal{O}(h^{2L+2}) , \tag{4.4}$$

where $D^s\, y_j \equiv y_j^{(s)}$.

Taking kth (operator) powers in (4.4) we obtain

$$(D_+D_-)^k\, y_j = [\, D^2 + \sum_{\nu=1}^{L} \frac{2D^{2\nu+2}}{(2\nu+2)!}\, h^{2\nu}]^k\, y_j + \mathcal{O}(h^{2L+2})$$

or

$$(D_+D_-)^k\, y_j = D^\mu\, y_j + \sum_{\nu=1}^{L} \alpha^e_{\nu\mu}\, h^{2\nu}\, y^{(2\nu+\mu)} + \mathcal{O}(h^{2L+2}) . \tag{4.5}$$

Thus, the exact rational coefficients $\alpha^e_{\nu\mu}$ are obtained through MACSYMA which very simply computes symbolically the kth power of the expansion, reorganizes it in powers of h^2 and drops the terms of order higher than $2L+2$.

Once the asymptotic expansion for the local error is available, the deferred correction procedure calls for approximating segments of this expansion to increasing orders in h. In the past we have used fast Vandermonde solvers to generate the weights for the necessary difference formulae (cf. Pereyra [1973]). This turns out to be quite an expensive procedure in this application and we prefer to use additional symbolic manipulation in order to generate a priori (and exactly) all the necessary weights and store them for later use.

The procedure is explained in detail and quite extensive tables of coefficients are presented in Keller and Pereyra [1977b].

At the qth step of the deferred correction procedure we will need to approximate q terms of the local truncation error to order h^{2q}. We denote by $\{u_j^{(q-1)}\}$ the (q-1)-corrected discrete solution which is accurate to order h^{2q}. The correction operator is

$$S_q(u_j^{(q-1)}) = \sum_{s=-(m+q)}^{m+q} \sum_{\nu=1}^{q} \Big[\sum_{\mu=\max(0,\,|s|-q)}^{m} \omega_s^{2(\mu+\nu),\,\tilde{q},\,e}\, \alpha^e_{\nu,\,2\mu}\, A_{2\mu}(t_j) h^{-2\mu}$$

$$+ \sum_{\mu=\max(0,\,|s|-q-1)}^{m-1} \omega_s^{2(\mu+\nu)+1,\,\tilde{q},\,e}\, \alpha^e_{\nu,\,2\mu+1}\, h^{-(2\mu+1)}\, A_{2\mu+1}(t_j)\Big]\, u_{j+s}^{(q-1)}$$

where $\tilde{q} = q-\nu+1$, and the weights $\omega_s^{\ell,q}$ are coefficients in the expansions of $D^r_{2q,\,e}$, the $O(h^{2q})$ approximation to D^r. Both the approximations and their expansions are obtained with MACSYMA.

The qth corrected value is then obtained by solving the linear system

$$\mathcal{A} u = f + S_q(u^{(q-1)}) ,$$

and

$$u^{(q)} - y = \mathcal{O}(h^{2q+2}) .$$

REFERENCES

de Boor, C. and R. Weiss, "SOLVEBLOK: A package for solving almost block diagonal linear systems, with applications to spline approximation and the numerical solution of ordinary differential equations," MRC Techn. Rep. No. 1625, Madison, Wisc. (1976).

Fox, L., Numerical Solution of Two-Point Boundary Value Problems, Clarendon Press, Oxford (1957).

Keller, H. B., "Accurate difference methods for nonlinear two-point boundary value problems," SIAM J. Numer. Anal. 11, 305-320 (1974).

Keller, H. B. and V. Pereyra, "Difference methods and deferred corrections for ordinary boundary value problems," To appear.

Keller, H. B. and V. Pereyra, "Symbolic generation of finite difference formulae," To appear.

Kreiss, H.-O., "Difference approximations for boundary and eigenvalue problems for ordinary differential equations," Math. Comp. 26, 605-624 (1972).

Lentini, M. and V. Pereyra, "PASVA2-Two point boundary value problem solver for nonlinear first order systems," Lawrence Berkeley Lab. program documentation report (1975).

Lentini, M. and V. Pereyra, "An adaptive finite difference solver for nonlinear two-point boundary problems with mild boundary layers," SIAM J. Numer. Anal. 14, 91-111 (1977).

Pereyra, V., "Iterated deferred corrections for nonlinear operator equations," Numer. Math. 10, 316-323 (1967).

Pereyra, V. High Order Finite Difference Solution of Differential Equations. STAN-CS-73-348. Stanford Univ., California (1973).

Pereyra, V. and W. H. K. Lee, "Solving two-point seismic ray-tracing problems in a heterogeneous medium. Part I: A general numerical method based on adaptive finite differences," To appear.

MACSYMA Reference Manual. The Math Lab. Group, Project MAC. MIT, Boston (1975).

Pereyra, V., W. Proskurowski and O. Widlund, "High order fast Laplace solvers for the Dirichlet problem on general regions," Math. Comp. 31, 1-16 (1977).

A FAST ALGORITHM FOR NONLINEARLY CONSTRAINED OPTIMIZATION CALCULATIONS

M.J.D. Powell

1. Introduction

An algorithm for solving the general constrained optimization problem is presented that combines the advantages of variable metric methods for unconstrained optimization calculations with the fast convergence of Newton's method for solving nonlinear equations. It is based on the work of Biggs (1975) and Han (1975, 1976). The given method is very similar to the one suggested in Section 5 of Powell (1976). The main progress that has been made since that paper was written is that through calculation and analysis the understanding of that method has increased.

The purpose of the algorithm is to calculate the least value of a real function $F(\underline{x})$, where $\underline{x}$ is a vector of n real variables, subject to the constraints

$$\left.\begin{array}{l} c_i(\underline{x}) = 0, \quad i = 1, 2, \ldots, m' \\ c_i(\underline{x}) \geqslant 0, \quad i = m'+1, m'+2, \ldots, m \end{array}\right\} \qquad (1.1)$$

on the value of $\underline{x}$. We suppose that the objective and constraint functions are differentiable and that first derivatives can be calculated. We let $\underline{g}(\underline{x})$ be the gradient vector

$$\underline{g} = \underline{\nabla} F(\underline{x}) \qquad (1.2)$$

and we let N be the matrix whose columns are the normals, $\underline{\nabla} c_i$, of the "active constraints".

The given algorithm is a "variable metric method for constrained optimization". The meaning of this term is explained in Section 2. Methods of this type require a positive definite matrix of dimension n to be revised as the calculation proceeds and they require some step-length controls to force convergence from poor starting approximations. Suitable techniques are described in Sections 3 and 4 and thus the recommended algorithm is defined. It is applied to some well-known examples in Section 5 and the numerical results are excellent. It seems to be usual for our algorithm to require less than half of the amount of work that is done by the best of the other published algorithms for constrained optimization. A theoretical analysis of some of the convergence properties of our method is reported elsewhere (Powell, 1977).

2. Variable metric methods for constrained optimization

Variable metric methods have been used successfully for many years for unconstrained optimization calculations. A good survey of their properties in the unconstrained case is given by Dennis and Moré (1977). Each iteration begins with a starting point $\underline{x}$ in the space of the variables at which the gradient (1.2) is calculated. A positive definite matrix, B say, which is often set to the unit matrix initially, defines the current metric. The vector $\underline{d}$ that minimizes the quadratic function

$$Q(\underline{d}) = F(\underline{x}) + \underline{d}^T \underline{g} + \tfrac{1}{2} \underline{d}^T B\underline{d} \tag{2.1}$$

is calculated. It is used as a search direction in the space of the variables, $\underline{x}$ being replaced by the vector

$$\underline{x}^* = \underline{x} + \alpha \underline{d}, \tag{2.2}$$

where α is a positive multiplier whose value depends on the form of the function of one variable

$$\phi(\alpha) = F(\underline{x} + \alpha \underline{d}). \tag{2.3}$$

The matrix B is revised, using the gradients $\underline{g}$ and $\underline{g}^* = \underline{\nabla}F(\underline{x}^*)$. Then another iteration is begun. We follow Han (1976) in seeking extensions to this method in order to take account of constraints on the variables.

When there are just n equality constraints on $\underline{x}$ and when the matrix of constraint normals

$$N = \{ \underline{\nabla}c_1, \underline{\nabla}c_2, \ldots, \underline{\nabla}c_{m'} \} \tag{2.4}$$

has full rank, an obvious method of defining the search direction $\underline{d}$ is to satisfy the equations

$$N^T\underline{d} + \underline{c} = 0, \tag{2.5}$$

where the components of $\underline{c}$ are the constraint values $c_i(\underline{x})$ $(i = 1, 2, \ldots, m')$. This is just Newton's method for solving the constraint equations if we let α equal one in the definition (2.2). When m' is less than n the advantages of Newton's method and the variable metric algorithms for unconstrained optimization are combined by calculating $\underline{d}$ to minimize the function (2.1) subject to the linear conditions

$$\left.\begin{aligned} \underline{\nabla c}_i^T \underline{d} + c_i &= 0, \quad i = 1, 2, \ldots, m' \\ \underline{\nabla c}_i^T \underline{d} + c_i &\geqslant 0, \quad i = m'+1, m'+2, \ldots m \end{aligned}\right\}, \tag{2.6}$$

which is a quadratic programming problem. Provided that each matrix B is positive definite and that each new value of $\underline{x}$ has the form (2.2), we say that an algorithm that calculates $\underline{d}$ in this way is a "variable metric method for constrained optimization".

One of the main conclusions of this paper is that it is satisfactory to keep the matrix B positive definite. This point is easy to accept when there are no constraints because then it is usual for the second derivative matrix of $F(\underline{x})$ to be positive definite at the required minimum so B can be regarded as a second derivative approximation. However, when there are constraints it may not be possible to identify B in this way. For example, consider the simple problem of calculating the least value of the function of one variable $F(x_1) = -x_1^2$ subject to the equation $x_1 = 1$. In this case any positive definite matrix B is satisfactory although $F(\underline{x})$ has negative curvature. We see that it is unnecessary to add a penalty term to $F(\underline{x})$ in order to make the second derivative of the objective function positive, which is done in the augmented Lagrangian method (Fletcher, 1975). One of the important features of the given algorithm is that there are no penalty terms. Our numerical results and the convergence theory in Powell (1977) show that a superlinear rate of convergence is usually obtained, even when the true second derivative matrix of the "Lagrangian function" is indefinite. The reason for mentioning the

Lagrangian function is given in the next section.

Keeping B positive definite not only helps the quadratic programming calculation that provides $\underline{d}$ but it also allows a basic feature of variable metric methods to be retained. It is that, except for the initial choice of B, the given method is invariant under linear transformations of the variables. Thus the main steps of the algorithm do not require the user to scale the variables and the constraint functions carefully before starting the calculation. However, some preliminary scaling may be helpful to the subroutines for matrix calculations that are used by the algorithm. For the results reported in Section 5 the search directions $\underline{d}$ were calculated by the quadratic programming subroutine due to Fletcher (1970), but this subroutine is not entirely satisfactory because it does not take advantage of the fact that B is positive definite.

Another feature of the quadratic programming calculation that ought to be mentioned, even though it is not relevant to the examples of Section 5, is that it may happen that the linear conditions (2.6) on $\underline{d}$ are inconsistent. If this case occurs it does not necessarily imply that the nonlinear constraints (1.1) are inconsistent. Therefore we introduce an extra variable, ξ say, into the quadratic programming calculation and we replace the constraints (2.6) by the conditions

$$\left.\begin{array}{ll} \underline{\nabla} c_i^T \underline{d} + c_i \xi = 0, & i = 1, 2, \ldots, m' \\ \underline{\nabla} c_i^T \underline{d} + c_i \xi_i \geq 0, & i = m'+1, m'+2, \ldots, m \end{array}\right\}, \tag{2.7}$$

where ξ_i has the value

$$\left.\begin{array}{ll} \xi_i = 1, & c_i > 0 \\ \xi_i = \xi, & c_i \leq 0 \end{array}\right\}. \tag{2.8}$$

Thus we modify only the constraints that are not satisfied at the starting point of the iteration. We make ξ as large as possible subject to the condition $0 \leq \xi \leq 1$. Any freedom that remains in $\underline{d}$ is used to minimize the quadratic objective function (2.1). Usually $\xi = 1$, in which case the calculation is the same as before. Any positive value of ξ allows a helpful correction to $\underline{x}$ of the form (2.2). However, it may happen that the only feasible solution of the conditions (2.7) occurs when ξ and the components of $\underline{d}$ are zero. In this case no small change to $\underline{x}$ makes a first order improvement to the violations of the nonlinear constraints (1.1), so the algorithm finishes because it is assumed that the constraints are inconsistent.

3. The revision of B

In order to achieve superlinear convergence the matrix B has to include some second derivative information, which is gained by the method that is used for revising B. A remark in Section 2 suggests that second derivatives of the Lagrangian function are more relevant than second derivatives of $F(\underline{x})$. This is certainly the case, but the point is missed by some published algorithms for constrained optimization. The importance of the Lagrangian function is easy to see when all the constraints are equalities and $m' < n$.

In this case the required vector $\underline{x}$ satisfies the equality constraints and the equation

$$\underline{\nabla}F(\underline{x}) - \sum_{i=1}^{m'} \lambda_i \underline{\nabla}c_i(\underline{x}) = 0, \tag{3.1}$$

where $\lambda_i (i=1,2,\ldots,m')$ are the Lagrange parameters. These conditions provide $m'+n$ equations in $m'+n$ unknowns, so one way of viewing the convergence rate of our algorithm is to compare it with Newton's method for solving the equations. We note that the second derivative information that is required by Newton's method is contained in the second derivative matrix with respect to $\underline{x}$ of the function

$$\Phi(\underline{x}, \underline{\lambda}) = F(\underline{x}) - \underline{\lambda}^T \underline{c}(\underline{x}). \tag{3.2}$$

The point that is important to the present discussion is that second derivatives of $F(\underline{x})$ alone are not helpful, except in special cases, for example when all the constraints are linear.

Therefore our method for revising B depends on estimates of Lagrange parameters. It is suitable to let $\underline{\lambda}$ be the vector of Lagrange parameters at the solution of the quadratic programming problem that defines $\underline{d}$, so it can be calculated quite easily and it changes from iteration to iteration. Thus the components of $\underline{\lambda}$ that correspond to inactive inequality constraints become zero automatically. Experience with variable metric methods for unconstrained optimization suggests that B should be replaced by a matrix, B* say, that depends on B and on the difference in gradients

$$\underline{\gamma} = \underline{\nabla}_x \Phi(\underline{x} + \underline{\delta}, \underline{\lambda}) - \underline{\nabla}_x \Phi(\underline{x}, \underline{\lambda}), \tag{3.3}$$

where $\Phi(\underline{x}, \underline{\lambda})$ is the function (3.2) and where $\underline{\delta}$ is the change in variables

$$\underline{\delta} = \underline{x}^* - \underline{x}. \tag{3.4}$$

When there are no constraints it is possible to choose the step-length α in equation (2.2) so that the scalar product $\underline{\delta}^T\underline{\gamma}$ is positive. The methods that are used for revising B suppose that this has been done, but, when there are constraints, it can happen that $\underline{\delta}^T\underline{\gamma}$ is negative for all non-zero values of α. In this case the usual methods for revising B would fail to make the matrix B* positive definite. Therefore we replace $\underline{\gamma}$ by the vector of form

$$\underline{\eta} = \theta\underline{\gamma} + (1-\theta)\, B\underline{\delta}, \quad 0 \leq \theta \leq 1, \tag{3.5}$$

that is closest to $\underline{\gamma}$ subject to the condition

$$\underline{\delta}^T\underline{\eta} \geq 0.2\ \underline{\delta}^T B\underline{\delta}, \tag{3.6}$$

where the factor 0.2 was chosen empirically. Thus θ has the value

$$\theta = \begin{cases} 1, & \underline{\delta}^T\underline{\gamma} \geq 0.2\ \underline{\delta}^T B\underline{\delta}, \\ \dfrac{0.8\underline{\delta}^T B\underline{\delta}}{\underline{\delta}^T B\underline{\delta} - \underline{\delta}^T\underline{\gamma}}, & \underline{\delta}^T\underline{\gamma} < 0.2\ \underline{\delta}^T B\underline{\delta}, \end{cases} \tag{3.7}$$

(Powell, 1976). We may use $\underline{\eta}$ in place of $\underline{\gamma}$ in several of the formulae that are applied by unconstrained optimization algorithms for revising B. We prefer the BFGS formula

$$B^* = B - \frac{B\underline{\delta}\,\underline{\delta}^T B}{\underline{\delta}^T B\underline{\delta}} + \frac{\underline{\eta}\,\underline{\eta}^T}{\underline{\delta}^T\underline{\eta}}, \tag{3.8}$$

because it is very successful in unconstrained calculations, because positive definiteness of B and condition (3.6) imply that B* is positive definite and because it gives invariance under changes of scale of the variables.

Powell (1977) shows that this method of revising B can give superlinear convergence even when the second derivative matrix of the Lagrangian function, G say, is indefinite. The method of proof is based on a comparison between projections of B and G, the projection being into the space that is the intersection of the tangent hyperplanes to the active constraints. The analysis is complicated by the fact that each search direction $\underline{d}$ has a part, proportional to $\underline{c}(\underline{x})$, that cuts across the tangent planes. It suggests that it may be better to leave B unchanged unless the ratio $\|\underline{c}\| / \|\underline{d}\|$ is small. This idea was investigated but was found not to be worthwhile. One reason is that Han's(1976) work shows that there is no need for the modification when G is positive definite. Another reason comes from the fact that, if the curvature of the function (2.1) is small, then the solution of the quadratic programming problem that defines $\underline{d}$ is usually at a vertex. In this case a scale-invariant implementation of the idea leaves B unchanged. Hence the idea gives a bias against correcting B when its diagonal elements are small.

4. The step-length parameter

The step-length parameter α in equation (2.2) is extremely important because it is used to force convergence from poor starting approximations. However, the choice of step-length is complicated by the fact that, not only do we wish to reduce the objective function, but also we have to satisfy the constraints. This need led to penalty function methods that, instead of minimizing $F(\underline{x})$, minimize a function of the form

$$\Psi(\underline{x}) = F(\underline{x}) + P[\underline{c}(\underline{x})] , \qquad (4.1)$$

where $P[\underline{c}(\underline{x})]$ is zero when the constraints are satisfied and that is positive otherwise. Because algorithms for minimizing functions of several variables were applied directly to $\Psi(\underline{x})$ extensions were made in order that $\Psi(\underline{x})$ became differentiable, the most successful technique of this kind being the augmented Lagrangian method. However, Han (1975) shows that there is no need for differentiability if the only use of $\Psi(\underline{x})$ is to help the choice of the step-length parameter. Therefore we follow his advice and use an objective function of the form

$$\Psi(\underline{x}, \underline{\mu}) = F(\underline{x}) + \sum_{i=1}^{m'} \mu_i |c_i(\underline{x})| + \sum_{i=m'+1}^{m} \mu_i |\min[0, c_i(\underline{x})]|, \quad (4.2)$$

requiring that the value of α in equation (2.2) satisfies the condition

$$\Psi(\underline{x}^*, \underline{\mu}) < \Psi(\underline{x}, \underline{\mu}), \quad (4.3)$$

where the components of $\underline{\mu}$ are defined later.

Condition (4.3) can be obtained if the function

$$\phi(\alpha) = \Psi(\underline{x} + \alpha\underline{d}, \underline{\mu}), \quad (4.4)$$

which reduces to expression (2.3) when there are no constraints, decreases initially when α is made positive. Han (1975) proves that this happens if B is positive definite and if the inequalities

$$\mu_i \geqslant |\lambda_i|, \qquad i = 1, 2, \ldots, m, \quad (4.5)$$

hold, where, as in Section 3, $\underline{\lambda}$ is the vector of Lagrange parameters at the solution of the quadratic programming problem that defines $\underline{d}$. He also shows that, if $\underline{\mu}$ satisfies condition (4.5) on every iteration, then convergence to the required vector of variables can be obtained from remote starting approximations. Therefore he suggests that $\underline{\mu}$ be a sufficiently large constant vector.

However, Powell (1976) notes that a constant vector $\underline{\mu}$ that satisfies condition (4.5) on every iteration may be inefficient, because it can happen that on most iterations $\underline{\mu}$ is much larger than necessary, in which case too much weight is given to satisfying the constraints on the variables. This situation occurs when the initial choice of B is too large, because there is a contribution to $\underline{\lambda}$ that is proportional to B. Therefore Powell (1976) suggests letting $|\underline{\mu}|$ be equal to $|\underline{\lambda}|$ on each iteration. However, further numerical experimentation shows that it can be advantageous to include positive contributions in the function (4.2) from some of the inequality constraints that are inactive at the solution of the quadratic programming problem that gives $\underline{\lambda}$. Therefore the following value of $\underline{\mu}$ is used in the algorithm that we recommend. On the first iteration we let $\mu_i = |\lambda_i|$ $(i = 1, 2, \ldots, m)$. On the other iterations we apply the formula

$$\mu_i = \max\left[|\lambda_i|, \tfrac{1}{2}(\bar{\mu}_i + |\lambda_i|)\right], \quad i = 1, 2, \ldots, m, \quad (4.6)$$

where $\bar{\mu}_i$ is the value of μ_i that was used on the previous iteration.

Because $\underline{\mu}$ changes on each iteration, Han's (1975) global convergence theorems do not apply. Therefore, as was the case when variable metric algorithms for unconstrained optimization were proposed, we cannot guarantee the success of the given method. The present Fortran program includes the following trap to catch some cyclic behaviour of the iterations. On each iteration and for each value of i $(i = 1, 2, \ldots, m)$ we let $\hat{\mu}_i$ be the greatest value of $|\lambda_i|$ that has been calculated. We note the minimum value of $\Psi(\underline{x}, \hat{\underline{\mu}})$ that occurs during each sequence of iterations for which $\hat{\underline{\mu}}$ remains constant. An error return is made if there is a run of five iterations where $\hat{\underline{\mu}}$ remains constant and the minimum value of $\Psi(\underline{x}, \hat{\underline{\mu}})$ does not decrease. This error return has never occurred. All the numerical calculations that have been tried suggest that the algorithm does converge satisfactorily from poor starting approximations.

The procedure for choosing α is as follows. It depends on a number Δ that is usually the derivative $\phi'(0)$, where $\phi(\alpha)$ is the function (4.4). We build a sequence $\alpha_k (k = 0, 1, 2, \ldots)$ until it gives a suitable value of α. The first term in the sequence is $\alpha_0 = 1$ and, for $k \geqslant 1$, the value of α_k depends on the quadratic approximation to $\phi(\alpha)$, $\phi_k(\alpha)$ say, that is defined by the equations

$$\left.\begin{aligned} \phi_k(0) &= \phi(0) \\ \phi_k'(0) &= \Delta \\ \phi_k(\alpha_{k-1}) &= \phi(\alpha_{k-1}) \end{aligned}\right\}. \tag{4.7}$$

We let α_k be the greater of $0.1\alpha_{k-1}$ and the value of α that minimizes $\phi_k(\alpha)$. For each term in the sequence we test the condition

$$\phi(\alpha_k) \leqslant \phi(0) + 0.1\,\alpha_k \Delta \tag{4.8}$$

and we set the step-length to α_k as soon as this inequality is satisfied. Methods of this type are used frequently in algorithms for unconstrained optimization.

However, it should be noted that, because of the derivative discontinuities in the function (4.2), the value of Δ is not always equal to $\phi'(0)$. We define Δ differently if a derivative discontinuity is expected to occur in $\phi(\alpha)$ for $0 < \alpha < 1$, for then the gradient $\phi'(0)$ may give misleading information about $\phi(\alpha)$ on the interval $[0,1]$, particularly if the discontinuity occurs near $\alpha = 0$. In all cases we set Δ to the difference $[\phi(1) - \phi(0)]$ that would occur if the functions $F(\underline{x})$ and $c_i(\underline{x})$ $(i = 1, 2, \ldots, m)$ were all linear. This difference is easy to compute because the gradients $\underline{g}$ and $\underline{\nabla} c_i$ $(i = 1, 2, \ldots, m)$ are known at the starting point of the iteration.

The numerical results of the next section show that on nearly every iteration the step-length α has the value one.

5. Numerical Results

The given algorithm has been applied to several test problems, including some where the nonlinear constraints define surfaces that include ripples. We report here on experience with three of Colville's (1968) test problems, with the Post Office Parcel problem (Rosenbrock,1960) and with a problem suggested by Powell(1969).

Colville's third problem includes five variables and sixteen constraints, six of them being nonlinear. It is the easiest of the examples because five constraints are active at the solution, which are identified on every iteration by the quadratic programming calculation that defines $\underline{d}$. Thus the algorithm reduces the problem to the solution of five equations in five unknowns and only two of the equations are nonlinear. Throughout the calculation a step-length of one is used so Newton's method is being applied to solve the equations. What is surprising in this case is not that our algorithm is fast but that the algorithms reported by Colville are so slow.

Colville's first problem is more interesting because, given the starting point that he recommends, it is not obvious which of the fifteen constraints are active at the solution. Our algorithm identifies these constraints successfully on the second iteration. There are four, they are all linear, while the number of variables is only five. Hence at an early stage the problem is reduced to the minimization of a function of only one variable. Thus fast convergence is obtained.

In Colville's second problem there are fifteen variables and twenty constraints. Using his infeasible starting point we find the final set of active constraints on the tenth iteration. There are eleven active constraints, eight of them being nonlinear. Hence in this example the algorithm does have to combine Newton's method for satisfying the constraint conditions with a minimization calculation to take up the remaining freedom in the variables.

The post office parcel problem has three variables and seven linear constraints, but only one constraint is active at the solution, which is identified on the second iteration when the standard starting point (10, 10, 10) is used. Because of symmetry in the second and third variables the problem is really reduced to the minimization of a function of one variable, so again the rate of convergence is rapid.

Similarly in Powell's problem, which has five variables and three nonlinear equality constraints, there is symmetry between the last two variables. Hence again the minimization part of the calculation has to take up only one degree of freedom. The nonlinearity of the constraints makes this calculation more testing than the post office parcel problem.

These remarks emphasize an important difference between our algorithm and methods that minimize a function of n variables on every iteration. It is that we are using active constraints to reduce the number of degrees of freedom in the minimization calculation. Similar savings are made by reduced gradient methods (see Sargent, 1974, for example), but they keep the constraint violations small throughout the calculation, which is analogous to choosing large values of μ_i $(i=1,2,\ldots,m)$ in expression (4.2).

A comparison of the present algorithm with some other methods is shown in Table 1. The given figures are the number of function and gradient evaluations that are required to solve each constrained minimization problem, except that the figures in brackets are the number of iterations. The five test problems have been mentioned already. The initial values of $\underline{x}$ for the Colville problems are the feasible starting points that he suggests, while on the last two problems the starting points are (10, 10, 10) and (-2, 2, 2, -1, -1) respectively. The first three columns of figures are taken from Colville (1968), Biggs (1972) and Fletcher (1975). In the case of our algorithm we suppose that the solution is found when all the components of $\underline{x}$ are correct to five significant decimal digits.

Colville's (1968) report compares most of the algorithms for constrained minimization calculations that were available in 1968. For each of his test problems he gives results for at least ten methods. We quote the smallest of his figures, even though the three given numbers are obtained by three different algorithms.

The results due to Biggs (1972) were calculated by his REQP Fortran program which is similar to the method that we prefer. Two important differences are the way he approaches constraint boundaries and his use of the objective function $F(\underline{x})$ instead of the Lagrangian function (3.2) in order to revise the matrix that we call B. He now uses the Lagrangian function to revise B (Biggs, 1975), but this change influences the numerical results only when some of the active constraints are nonlinear.

Fletcher (1975) studies three versions of the augmented Lagrangian method. He gives figures for each version and, as in the case of the Colville results, our table shows the figures that are most favourable to Fletcher's work on each test problem. It is incorrect to infer from the table that some of Colville's algorithms are superior to the augmented Lagrangian method because most of the early algorithms are rather inconsistent, both in the amount of work and in the final accuracy.

We do, however, claim that the table shows that the algorithm described in this paper is a very good method of solving constrained optimization calculations with nonlinear constraints. It can be programmed in an afternoon if one has a quadratic programming subroutine available to calculate $\underline{d}$ and $\underline{\lambda}$. It is usually unneccessary to use the form (2.7) of the linear approximations to the constraints instead of the form (2.6). Also the device that depends on $\hat{\underline{\mu}}$, described in the paragraph that follows equation (4.6), seems to be unnecessary. Such a program is usually satisfactory for small calculations and for calculations where most of the computer time is spent on function and gradient evaluations. In other cases, however, the matrix calculations of the algorithm may dominate the running time of the program. Therefore the quadratic programming part should be solved by an algorithm that takes advantage of structure, such as the form of the change to B that occurs on each iteration.

TABLE I

Comparison of Algorithms

PROBLEM	COLVILLE	BIGGS	FLETCHER	PRESENT
COLVILLE 1	13	8	39 (4)	6 (4)
COLVILLE 2	112	47	149 (3)	17 (16)
COLVILLE 3	23	10	64 (5)	3 (2)
POP	—	11	30 (4)	7 (5)
POWELL	—	—	37 (5)	7 (6)

References

Biggs, M.C. (1972) "Constrained minimization using recursive equality quadratic programming" in Numerical methods for nonlinear optimization, ed. F.A. Lootsma, Academic Press (London).

Biggs, M.C. (1975) "Constrained minimization using recursive quadratic programming: some alternative subproblem formulations" in Towards global optimization, eds. L.C.W. Dixon and G.P. Szegö, North-Holland Publishing Co. (Amsterdam).

Colville, A.R. (1968) "A comparative study on nonlinear programming codes", Report No. 320-2949 (IBM New York Scientific Center).

Dennis, J.E. and Moré, J. (1977) "Quasi-Newton methods, motivation and theory", SIAM Review, Vol. 19, pp. 46-89.

Fletcher, R. (1970) "A Fortran subroutine for quadratic programming", Report No. R6370 (A.E.R.E., Harwell).

Fletcher, R. (1975) "An ideal penalty function for constrained optimization", J. Inst. Maths. Applics., Vol 15, pp. 319-342.

Han, S-P. (1975) "A globally convergent method for nonlinear programming", Report No. 75-257 (Dept. of Computer Science, Cornell University).

Han, S-P (1976) "Superlinearly convergent variable metric algorithms for general nonlinear programming problems", Mathematical Programming, Vol. 11, pp. 263-282.

Powell, M.J.D. (1969) "A method for nonlinear constraints in minimization problems" in Optimization, ed. R. Fletcher, Academic Press (London).

Powell, M.J.D. (1976) "Algorithms for nonlinear constraints that use Lagrangian functions", presented at the Ninth International Symposium on Mathematical Programming, Budapest.

Powell, M.J.D. (1977) "The convergence of variable metric methods for nonlinearly constrained optimization calculations", presented at Nonlinear Programming Symposium 3, Madison, Wisconsin.

Rosenbrock, H.H. (1960) "An automatic method for finding the greatest or the least value of a function", Computer Journal, Vol. 3, pp. 175-184.

Sargent, R.W.H. (1974) "Reduced-gradient and projection methods for nonlinear programming" in Numerical methods for constrained optimization, eds. P.E. Gill and W. Murray, Academic Press (London).

THE DECOMPOSITION OF SYSTEMS OF PROCEDURES AND ALGEBRAIC EQUATIONS

R.W.H. Sargent

1. INTRODUCTION

We are here concerned with the decomposition of large-scale computing problems which have a network structure. The problem arises in many different fields, such as the design or simulation of chemical processes, electrical circuit design, or the study of distribution networks of various kinds; the techniques can even be applied to the analysis of flow-diagrams for computer programmes. Contributions to solution of the problem are similarly widely spread, but no attempt will be made here to give a systematic review of the literature or to trace the history of development of ideas. A good review has been given by Duff (1976).

We shall first consider systems of interlinked procedures or subroutines, then sparse systems of algebraic equations, and finally present an algorithm for decomposing mixed systems.

2. SYSTEMS OF PROCEDURES

A network of interlinked procedures can be represented by a directed graph, as shown in Fig. 1, where the nodes represent the procedures and the arcs streams of information flowing between them. The procedures process their input information to produce output information and cannot be executed until all their input information is provided.

Arcs with no generating node, such as A, D, J, S, V in Fig. 1, are given information or data, so that the procedure corresponding to node 1 can be executed immediately; this then completes the input information to node 4 which can be executed in turn. However, the complete network cannot be evaluated by a simple sequence of computations because of the existence of loops or cycles of information, represented in Fig. 1 by arcs M and N mutually linking nodes 3 and 9, and by L, R, Q, P, T mutually linking nodes 5, 6, 7 and 8. To evaluate such a group of nodes linked by a cycle an iterative scheme is necessary, and to start the calculation we must assume the information in one or more arcs so that the cycle is broken. For example, we may assume information represented by arc M in Fig. 1, thus enabling node 3 to be executed, followed by 9, which yields new values for arc M. If x denotes the set of variables defining M, and $f(x)$ is the result of computing the new values via execution of 3 and 9, the problem to be solved is to find a fixed point of the mapping:

$$x = f(x) \tag{1}$$

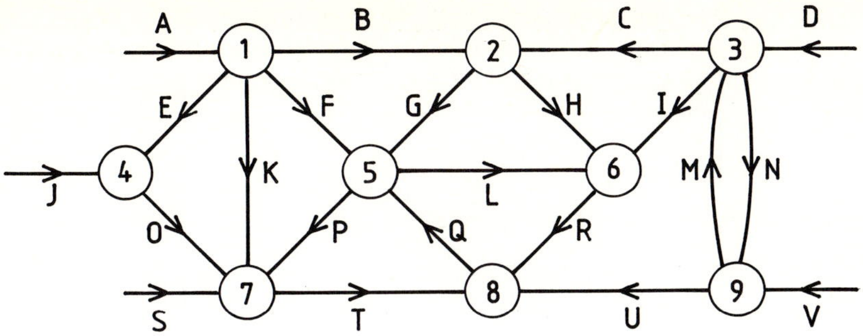

Figure 1 Directed Graph

Thus the first stage of decomposition is to order the nodes into a computable sequence of groups, each group consisting of either a single node, or a set of nodes linked by one or more cycles. Such groups of nodes are called the "strong components" of the graph, and we shall refer to such a sequence as a "partition" of the graph. For Fig. 1 it is easily seen by inspection that a possible partition is

$$1,\ 4,\ (3,\ 9),\ 2,\ (5,\ 6,\ 7,\ 8) \tag{2}$$

Clearly (3, 9) can also be placed first, or between 1 and 4, in this sequence. The choice between these alternatives will depend on the relative amounts of information generated, and the consequent need for temporary relegation of information to backing store.

The second stage is the decomposition of the linked groups of nodes by opening or "tearing" the cycles, and the set of arcs representing the guessed information is known as an "essential arc set" or a "tear set". In Fig. 1 suitable tear sets for the group (5, 6, 7, 8) are either (P, R) or simply (Q).

2.1 Partitioning Algorithms

A directed graph can be represented by its adjacency matrix (A), which is a square matrix with a row and column corresponding to each node of the graph and elements a_{ij} defined by

$$a_{ij} = \begin{cases} 1 \text{ if there is a directed arc from j to i,} \\ 0 \text{ otherwise} \end{cases} \tag{3}$$

Thus the units in row i indicate the source nodes of information required to execute node i. The adjacency matrix for the graph of Fig. 1 is given in Fig. 2a.

A re-ordering of the nodes corresponds to a transformation PAP^T of the adjacency matrix, where P is a permutation matrix, and if the re-ordering results in a

Figure 2 Adjacency Matrix

(a)

	1	2	3	4	5	6	7	8	9
1									
2	1		1						
3									1
4	1								
5	1	1						1	
6		1	1		1				
7	1			1	1				
8						1	1		1
9		1							

(b)

	1	4	3	9	2	5	6	7	8
1									
4	1								
3				1					
9			1						
2	1		1						
5	1				1				1
6			1		1	1			
7	1	1				1			
8				1			1	1	

computable sequence the transformed adjacency matrix is in block triangular form, the blocks corresponding to the strong components of the graph. This is illustrated in Fig. 2b for the permutation corresponding to the computable sequence given in (2).

The reachability matrix, R, of a graph is also a binary matrix with elements r_{ij} defined by:

$$r_{ij} = \begin{cases} 1 \text{ if there is a directed path from node j to node i,} \\ 0 \text{ otherwise} \end{cases} \quad (4)$$

Harary (1960, 1962a) has shown that the reachability matrix can be obtained from the adjacency matrix by the formula:

$$R = (A+I)^{n-1} \tag{5}$$

where the operations are carried out in Boolean arithmetic, and gave an algorithm for finding a partition of the graph based on the properties of R. Baker (1962), Warshall (1962) and Mah (1974) all give shorter methods of obtaining R than the direct use of (5). Ponstein (1966) exploited a different approach by Harary (1962b) based on the properties of the so-called "variable adjacency-matrix" (obtained from A by replacing unit elements by undetermined variables), giving an algorithm which finds a partition and all simple cycles of the graph.

It turns out that it is much quicker to use a direct algorithm which traces paths and cycles on the graph, and the first algorithm of this type was given by Sargent and Westerberg (1964). However, the efficiency of their algorithm depended crucially on the list-processing package of Cooper and Whitfield (1962) used in the computer implementation, and more recently Tarjan (1972) has published a similar algorithm using a much simpler list structure. A discussion of these two algorithms is given by Duff and Reid (1976), who also give a Fortran listing of Tarjan's algorithm. Again these algorithms give the simple cycles of the graph as a by-product.

2.2 Tearing Algorithms

Algorithms for tearing are most easily discussed in terms of the cycle-arc incidence matrix C, with columns corresponding to arcs, rows corresponding to the simple cycles of the graph, and elements c_{ij} defined by:

$$c_{ij} = \begin{cases} 1 \text{ if arc } j \text{ is in cycle } i \\ 0 \text{ otherwise} \end{cases} \tag{6}$$

Figure 3 Cycle-Arc Incidence Matrix

Arcs	L	M	N	P	Q	R	T
Cycles							
1	1				1	1	
2				1	1		1
3		1	1				
Parameters	3	4	7	2	6	4	5

The cycle-arc matrix for the graph of Fig. 1 is given in Fig. 3, where an additional row has been added to indicate the number of parameters associated with each arc.

A tear set is a set of arcs such that each is in at least one cycle and each cycle contains at least one of the arcs. If each arc j is associated with a cost or weight w_j, and we wish to find the tear set with the minimum total cost, the problem may be formulated as:

$$\left.\begin{array}{lll} \text{minimize} & \sum_{j=1}^{n} w_j x_j \;, & \\ \text{subject to} & \sum_{j=1}^{n} c_{ij} x_j \geq 1 \;, & i = 1,2,\ldots m, \\ \text{and} & x_j = 0 \text{ or } 1 \;, & j = 1,2,\ldots n. \end{array}\right\} \quad (7)$$

This is a standard set-covering problem and efficient algorithms are available for its solution, as described by Garfinkel and Nemhauser (1972).

It is interesting to note that many of the reduction-rules available for reducing the size of a set-covering problem before solution have been independently discovered by those working specifically on the tearing problem (Sargent and Westerberg (1964), Lee and Rudd (1966), Christensen and Rudd (1969), Barkley and Motard (1972)), and some of these give additional reduction-rules. These rules are so effective that they often yield the solution directly, or at least leave a relatively small residual set-covering problem.

It is not of course necessary to form the cycle-arc incidence matrix explicitly, since the equivalent information can be stored in lists and is available from the partitioning stage.

The remaining question is an appropriate choice of the costs w_j. It has often been assumed that the greatest simplification will result from minimizing the number of arcs in the tear set, which of course corresponds to setting $w_j = 1$, all j, in (7). For Fig. 1 this yields optimum tear sets (Q, M) and (Q, N). Others have considered that the number of guessed parameters should be minimized, in which case w_j is taken as the number of parameters describing the information represented by arc j. For Fig. 1 this choice yields the optimum tear set (L, P, M).

Minimizing the number of guessed parameters minimizes the storage requirements, but it is not necessarily the best strategy for minimizing computation time. If direct successive substitution is used to solve (1), then both the conditioning and the convergence rate are improved by making the spectral norm of the Jacobian matrix $f_x(x)$ as small as possible. Although simple bounds on the norm might be used, as suggested by Westerberg and Edie (1971), this matrix would have to be generated by finite-difference approximation for each possible iteration loop, and it is almost certainly more profitable to adopt a simpler measure for choosing a tear set and use the computational effort saved on extra iterations.

Upadhye and Grens (1975) give an interesting analysis of the problem which leads to a simple heuristic rule for choosing the costs w_j. They base their analysis on the use of direct substitution, and start by pointing out that the convergence properties of a loop of procedures are independent of the point at which the loop is torn for initialization. Thus it is possible to define families of tear sets in which all members of the same family have equivalent convergence properties (corresponding to tearing the cycles at different points), and each family is characterized by the number of times each cycle is torn. They then show that for any directed graph there is at least one family of non-redundant tear sets (i.e. sets which tear no cycle more than once), and that if a family contains a redundant tear set then it also contains a tear set in which at least one arc is torn more than once.

Their empirical results on a large number of problems in the field of chemical process simulation indicate that convergence is in general much slower for iteration schemes involving such double-tearing of arcs. These results are supported by the observation that for linear systems where $f_x(x)$ is either diagonal or non-negative it can be shown that double-tearing must increase its spectral norm. Thus they conclude that one should seek a tear set from a non-redundant family. Such a tear set must exist, and is obtained by setting

$$w_j = \sum_{i=1}^{m} c_{ij} \quad , \quad j = 1,2,\ldots n \tag{8}$$

in problem (7). For Fig. 1 this rule shows the equivalence of the tear sets (Q, M), (Q, N), (L, P, M), (L, T, N) etc.

Clearly the equivalence of tear sets within these families breaks down if a more sophisticated solution procedure is used, but the approach is interesting and further work of this type could well pay dividends.

3. SYSTEMS OF ALGEBRAIC EQUATIONS

3.1 Partitioning and Tearing

Consider the system:

$$\begin{aligned}
f_1(x_1, x_2, x_4) &= 0\\
f_2(x_1, x_3) &= 0\\
f_3(x_1, x_2, x_3, x_4, x_6) &= 0\\
f_4(x_1, x_3) &= 0\\
f_5(x_2, x_5, x_6) &= 0\\
f_6(x_3, x_5) &= 0
\end{aligned} \tag{9}$$

This system could be solved as a simultaneous set of equations, using for example Newton's method, or better a quasi-Newton or secant method (cf. Ortega and Rheinboldt (1970)). This would require storage of the 6 x 6 Jacobian matrix or an approximation to it.

However, we could start by solving equations (2) and (4) for variables x_1 and x_3, then equation (6) for x_5, and finally equations (1), (5) and (3) for x_2, x_4, and x_6.

The maximum storage for this scheme is a 3 x 3 matrix, and the computations would be correspondingly reduced.

Analyses of decompositions of this sort are made more easily by use of the occurrence matrix M, which has a row for each equation, a column for each variable, and elements defined by

$$m_{ij} = \begin{cases} 1 \text{ if variable } j \text{ occurs in equation } i, \\ 0 \text{ otherwise} \end{cases} \tag{10}$$

The occurrence matrix for system (9) is given in Fig. 4a.

Figure 4 Occurrence Matrix

Equations \ Variables	1	2	3	4	5	6
1	1	1		1		
2	1		1			
3	1	1	1	1		1
4	1		1			
5		1			1	1
6			1		1	

(a)

Equations \ Variables	1	3	5	2	4	6
4	1	①				
2	①	1				
6		1	①			
1				1	①	
5			1	1		①
3	1	1		①	1	1

(b)

Re-ordering the equations and variables is equivalent to a transformation PMQ, where P and Q are permutation matrices, and the above decomposition implies a permutation to block triangular form, as shown in Fig. 4b for system (9). The blocks along the diagonal yield the structure of the Jacobian matrices of the subsystems, indicating the positions of non-zero elements in these matrices. It is possible that these subsystems can usefully be further decomposed by tearing. Thus system (9) can be decomposed into a sequence of one-variable problems by guessing variables x_1 and x_2; the residual of equation (2) can be used to correct the estimate of x_1, and that of equation (3) to correct x_2, each correction again forming a one-variable outer loop of iterations. More generally, the guessed (or torn) variables form one or more outer subsystems within each of which is nested a sequence of subsystems, and the extent of decomposition is arbitrary.

Clearly there is a close analogy between decompositions of systems of equations and systems of procedures, but there are important differences. In contrast to a procedure, an equation has no natural direction of computation; it may be easier to solve a given equation for certain variables, but in principle an equation may be solved for any of the variables occurring in it. However, once we have made a choice, the equation becomes a procedure for computing the chosen variable as "output" from the remaining variables as "inputs". We can then represent the resulting system of procedures by a directed graph whose nodes are the equations or output variables, with directed arcs indicating the successive substitutions of the variables.

We can permute the columns of the occurrence matrix to bring the chosen output variables onto the diagonal, and if these diagonal elements are then deleted the resulting matrix is the adjacency matrix of this associated graph:

$$A = M\overline{Q} - I, \tag{11}$$

where $\overline{Q}$ is a permutation matrix. Clearly the partitioning of this graph reduces A to block triangular form by a symmetric permutation PAP^T, and since such a permutation does not affect the diagonal elements the matrix $PM\overline{Q}P^T$ is also block triangular. Thus the equations can be partitioned using the techniques given in the previous section.

However, tearing variables in a system of equations corresponds to removing nodes from the associated graph, rather than arcs as discussed previously. This is still a set-covering problem of the form of equation (7), where the matrix C is now the cycle-node incidence matrix for the graph. Strodiot and Toint (1976) make the interesting alternative proposal of applying a Gauss-Seidel type of iteration procedure to the whole system, repetitively solving the equations for their output variables in sequence using the latest iterates of their input variables. On the grounds of minimizing the "feed-back" in the system, they choose the sequence to minimize the number of non-zero elements above the diagonal in the matrix $M\overline{Q}$, which is precisely the choice of a minimum essential arc set in the associated graph.

3.2 Feasible Output Sets

If the system of equations is soluble and non-redundant there must be at least one possible assignment of output variables to the equations. Such a feasible "output set" is not in general unique, but finding one is not a trivial problem. The problem is best understood by representing the system as a bipartite graph, with one set of nodes corresponding to the variables and the other set corresponding to the equations; if variable j occurs in equation i an arc is drawn joining nodes i and j. The graph for system (9) is given in Fig. 5.

Now a "matching" in a graph is a set of arcs such that no two arcs in the set are incident to the same node, and a "maximum matching" is a matching of maximum cardinality. Thus a feasible output set corresponds to a maximum matching in the bipartite graph, of cardinality equal to the number of variables or equations; if the maximum matching is of lower cardinality, there is no feasible output set and the equations are either inconsistent or contain redundancies.

Figure 5 Bipartite Graph

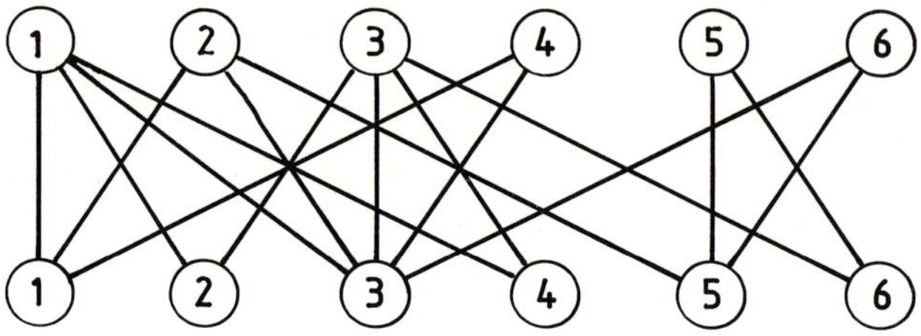

Finding an optimum maximum matching is a set-partitioning problem:

$$\left.\begin{aligned} &\text{Maximize} \quad \sum_j w_j x_j \\ &\text{subject to} \quad \sum_j c_{ij} x_j = 1 \quad , \quad i = 1,2,\ldots \\ &\text{and} \quad x_j = 0 \text{ or } 1, \; j = 1,2,\ldots \end{aligned}\right) \qquad (12)$$

where the matrix C (with elements c_{ij}) is here the node-arc incidence matrix for the bipartite graph. This matrix has only two non-zero elements in each column (since an arc joins only two nodes), and Edmonds and Johnson (1970) give an especially efficient algorithm to solve it. If one is content with any maximum matching, not necessarily optimal, then an even faster algorithm is available - see Edmonds (1965), Garfinkel and Nemhauser (1972), Hopcroft and Karp (1973).

Again the choice of output set affects the conditioning and rate of convergence, and this should be reflected in the values or weights w_j in (12). Strodiot and Toint (1976) suggest a heuristic allocation of weights based on a sensitivity analysis of each equation for the initial variable estimates, but give no details. Westerberg and Edie (1971) consider solution by successive substitution and recommend maximization of the absolute value of the product of diagonal elements of the iteration Jacobian; this of course destroys the special structure of problem (12) and they propose an implicit enumeration algorithm. It seems very doubtful that the work involved in this approach is justified, since the measure of optimality is still very crude.

3.3 Further Properties

Steward (1962, 1965) was the first to analyse systems of equations in this way, giving algorithms for finding a feasible output set, then the block triangular partition of the system, and finally the minimum tear set of variables for the reduction of each block to a sequence of one-variable problems. His algorithms provided a constructive proof that the partitioning was independent of the output set used to find it, but that the minimum tear sets did depend on the output set chosen. He also gives an algorithm for generating all feasible output sets from a first feasible set, which could be used as the basis for an implicit enumeration algorithm for the optimum decomposition of each block. However, even using the simplest cost function and the most efficient algorithms now available for each of the subproblems, this approach would still imply prohibitive computing requirements.

Further complications are suggested by the example given in Fig. 6, which gives an occurrence matrix already in partitioned form. Each block requires at least two torn variables to reduce it to a sequence of one-variable problems, giving four overall.

Figure 6 Occurrence Matrix

Equations \ Variables	1	2	3	4	5	6	7	8	9
1	1		(1)						
2	1	1		(1)					
3			1	1	(1)				
4		1	1		1				
5		1		1	1				
6						1	1	(1)	
7						1		1	(1)
8	1					1	1		1
9	1	1					1	1	1

For the first block a possible choice is to guess variables 1 and 2, computing 3, 4 and 5 in turn from equations 1, 2 and 3, then using the residuals of equations 4 and 5 to correct the guesses. Note however that this scheme does not use a feasible output set, since variable 1 does not occur in either of equations 4 or 5! In fact, it seems that there always exists a computational scheme based on a feasible output set which does realize the minimum number of torn variables, but when the costs w_j in (7) are not all equal it is clear that this set of computational schemes may not include the optimum scheme.

For the second block, variables 6 and 7 form a minimum tear set, corrected by the residuals of equations (8) and (9), and assigned output variables are ringed in the remaining equations. However, instead of starting by solving the first block, we could guess variables 6 and 7 and then solve the equations 6, 7, 8, 9, 1, 2, 3 in turn for the variables 8, 9, 1, 2, 3, 4, 5 respectively, finally using the residuals of equations 4 and 5 to correct the guesses. Hence the minimum tear set for the unpartitioned system is smaller than the union of the minimum tear sets for the partitioned blocks! Thus it is not necessarily optimal to partition first and then tear the individual blocks.

It is clear from these two observations that matters are not as simple as might at first appear. We need to examine more closely the nature of computational costs and the effect of partitioning and tearing on them, and then to develop an algorithm capable of considering these in relation to all valid computational schemes. We start by proving two theorems.

Theorem 1

The decomposition of a set of n equations in n variables into an ordered sequence of irreducible subsets is unique to within permutations of disjoint subsets.

Proof

Suppose we have a feasible sequence $\{A_j\}$, $j = 1,2,\ldots t$, of irreducible subsets, where subset A_j contains e_j equations, and suppose that there exists a second feasible sequence $\{B_i\}$, $i = 1, 2,\ldots,$ such that B_1 contains w equations, made up of m_j equations from subset A_j, $j = 1, 2,\ldots t$.

$$\text{Thus} \qquad w = \sum_{j=1}^{t} m_j \tag{13}$$

If the m_j equations from A_j contain n_j variables from this subset and p_j variables not otherwise introduced from preceding subsets $A_1, A_2,\ldots A_{j-1}$, then the number of variables in B_1 is

$$w = \sum_{j=1}^{t} (p_j+n_j) \quad , \quad p_1 \equiv 0 \tag{14}$$

Now there are three possibilities for m_j:

$$\left.\begin{array}{ll} \text{a)} & m_j = 0, \text{ whence of course } p_j = n_j = 0 \\ \text{b)} & 0 < m_j < e_j, \text{ whence } n_j > m_j \text{ since } A_j \text{ is irreducible} \\ \text{c)} & m_j = e_j, \text{ whence } n_j = e_j \text{ since } m_j \text{ is the entire block } A_j. \end{array}\right\} \tag{15}$$

Hence for all j we have $n_j \geqslant m_j$, and it follows from (13) and (14) that

$$w = \sum_{j=1}^{t} m_j = \sum_{j=1}^{t} (p_j+n_j) \geqslant \sum_{j=1}^{t} (p_j+m_j).$$

But m_j, n_j, p_j are all non-negative, so we must have equality throughout and hence

$$p_j = 0 \quad , \quad n_j = m_j \quad , \quad \text{all } j \tag{16}$$

From (15) and (16) we must have either $m_j = 0$ or $m_j = e_j$, and since B_1 is irreducible it must consist of only one subset A_j with $p_j = 0$, $m_j = n_j = e_j$. Again, since $p_j = 0$ it must be either A_1 or a succeeding disjoint block, and in the latter case the sequence will remain feasible if this is permuted to the first position.

Deleting this subset from both $\{A_j\}$ and $\{B_j\}$ leaves subsequences to which the same argument can be applied, and recursion of the process completes the proof. Q.E.D.

This result is not new and was indeed given by Steward (1962), but the above proof, which is a simplification of one given by Leigh (1973), does not introduce the irrelevant concept of an output set.

Theorem 2 cf. Leigh (1973)

The number of torn variables required for the complete reduction (to a sequence of one-variable problems) of a set of equations cannot be less than the minimum number for the complete reduction of any subset.

Proof

Consider a partition of the system into two subsets with t_1, t_2 torn variables respectively, and suppose that the number of torn variables for the unpartitioned system is t.

Now the variables of the second subset do not appear in equations of the first subset, so that in both the partitioned and unpartitioned cases they must either be torn or output variables of equations in the second subset. Since t_2 is the minimum number of torn variables for the latter, we have immediately $t \geq t_2$.

Now suppose that in the unpartitioned case p of the variables in the first subset are output variables of equations in the second subset. Then at least (t_1-p) variables of the first subset must be torn, for at best the p variables are members of the minimal tear set for the first subset. But the p equations in block 2 for which they are output variables are not available for determining variables of the second subset, so at least p variables of this subset must be torn.

$$\text{Thus} \qquad t \geq (t_1-p) + p = t_1$$

If either subset partitions further, the same argument can be applied to the successive subpartitions, and the result follows. Q.E.D.

The computational cost (c) of solving a set of n algebraic equations by tearing is given by

$$c = \sum_{j=1}^{N} \sum_{i=1}^{n} r_i N_{ij} \tag{17}$$

where N is the number of outer iterations on the torn variables, r_i is the computational cost of evaluating the residual of equation i, and N_{ij} is the number of iterations required to evaluate the output variable of equation i on the jth outer iteration. But for each of the last t equations (where t is the number of torn variables) the residual is used directly in the outer loop, so for these equations $N_{ij} = 1$ and equation (17) may be written

$$c = N \sum_{i=1}^{n} r_i + \sum_{j=1}^{N} \sum_{i=1}^{n-t} r_i (N_{ij}-1), \tag{18}$$

where $N_{ij} \geq 1$, all i and j.

If the output variables occur linearly in the one-variable problems we have $N_{ij} = 1$ for all i and j, and then under the assumption that N is a non-decreasing function of t the cost is minimized by minimizing the number of torn variables. More generally, this will be true if the outer loop iteration dominates the cost.

This raises the question of whether each nonlinear one-variable problem should be solved to high accuracy at each outer iteration, and in the limit it is possible to choose $N_{ij} = 1$ systematically, again leading to the above rule. If output variables were assigned from a feasible output set, it would then be possible to compute each of the torn variables from the residual equations in similar fashion, yielding a scheme analogous to that proposed by Strodiot and Toint (1976), but with a different criterion for choosing the order of computation. However, both these schemes are essentially direct substitution schemes and it will probably be more efficient to carry out some iteration on the one-variable problems and use a higher-order scheme on the outer iteration loop.

Now if we can assume that r_i, N_{ij} have approximately the same values for all i, and that N is a non-decreasing function of t, then clearly $\overline{N} = \sum_{j=1}^{N} (N_{ij}-1)$ is also a non-decreasing function of t, and (18) may be written

$$\begin{aligned} c &= NR + \overline{N}r(n-t) \\ \text{with} \quad R &= \sum_{i=1}^{n} r_i \end{aligned} \tag{19}$$

If now the system can be partitioned into two subsets, denoted by suffices 1 and 2, we have from (19) and Theorem 2, for minimum tear sets in each case:

$$n = n_1+n_2 \quad , \quad R = R_1+R_2 \quad , \quad t_1+t_2 \geq t \quad , \quad t \geq t_1 \quad , \quad t \geq t_2 \, . \tag{20}$$

From (19), (20) and the assumed monotonicity of N and $\overline{N}$ we then have

$$\begin{aligned} c - (c_1+c_2) &= \{NR+\overline{N}r(n-t)\} - \{N_1R_1+\overline{N}r(n_1-t_1)\} - \{N_2R_2+\overline{N}_2r(n_2-t_2)\} \\ &\geq N(R-R_1-R_2) + \overline{N}r(n-n_1-n_2) + \overline{N}r(t_1+t_2-t) \geq 0. \end{aligned} \tag{21}$$

Thus the cost for the partitioned system cannot exceed the unpartitioned cost, and since the same argument applies for further subpartitions the cost is minimized by partitioning into irreducible subsets. Of course, the assumptions leading to this result will often be invalidated, but they should at least indicate the general trend.

There remains the question of whether tearing is desirable at all, or whether one should not simply solve each irreducible subset in turn taking account of the sparsity.

For linear systems Duff (1976) reports results of several workers indicating that tearing and use of modification formulae is never more advantageous than direct solution of the system by factorization using sparse matrix techniques, but a similar general conclusion is not so obvious for nonlinear systems involving iteration. Clearly the outer loop system in the torn variables must be regarded as full, but the one-variable problems cause no fill-in, whereas some fill-in will occur if the whole system is solved as it stands; the balance of advantage depends on the proportion of torn variables and the sparsity structure, and so is likely to be very problem-dependent. Incomplete reduction by tearing will clearly be of advantage where judicious tearing of a few variables leaves a large linear subsystem, or where the one-variable problems generated are explicit in their output variables.

We end this section by considering rectangular systems, for in many applications there are more variables than equations and these form equality constraints in a nonlinear programme. It is then possible to use some or all of the equations to eliminate a corresponding number of variables from the optimization problem, and we have the flexibility of choosing these to facilitate the solution. If we have m equations in n variables we can choose $t \geqslant n-m$ "decision variables", which become the variables in a reduced nonlinear programme with $(t-n+m)$ equations remaining as equality constraints.

We can still represent the system as a bipartite graph, and a feasible output set is still a maximum matching of cardinality m. However, the irreducible subsets now depend on the choice of decision variables, and an arbitrary choice of these may turn out to be inconsistent since there may be subsystems which completely determine some of the variables.

The solution of the nonlinear programme determines the $(n-m)$ degrees of freedom in the system of equations, and we can imagine these as taken up by $(n-m)$ extra equations appended to the system. We can now partition this $n \times n$ system, and Theorem 1 assures us that the resulting partition is unique. If each of the appended hypothetical equations is assumed to contain all the n variables, then the partition can be arranged so that these are the last $(n-m)$ equations and they can be removed, leaving the original system in partitioned form with square blocks along the diagonal, except for the last block. Clearly $(n-m)$ of the decision variables must be taken from this last block, and for each additional decision variable an equation from the same block must be retained as an equality constraint. Fig. 7 gives a simple example of five equations in seven variables. It may well be possible to choose the $(n-m)$ free variables in the last block so that the remaining square subsystem partitions further, as for example by choosing variables 4 and 5 in Fig. 7.

4. AN ALGORITHM FOR MIXED SYSTEMS

We describe in this section an implicit enumeration algorithm for the decomposition of mixed systems of algebraic equations and procedures. It is essentially the algorithm developed by Leigh (1973) for systems of equations, with some minor changes and extensions to deal with procedures and the recognition of linear subsystems.

Figure 7 Rectangular System

Equations	Variables 1	2	3	4	5	6	7
1	1	1					
2	1	1					
3	1		1				
4		1		1	1	1	
5	1		1	1		1	1
-	1	1	1	1	1	1	1
-	1	1	1	1	1	1	1

Although later we consider the implications of more general costs, we start by considering complete reduction of the unpartitioned system assuming that the cost is proportional to the number of torn variables (the proportionality factor can be taken as unity without loss of generality). We also assume for the present that we have a rectangular system of m equations in $n \geqslant m$ variables.

First note that any permutation of the equations represents a feasible computational sequence. For any equation in a given sequence we guess all except one of the new variables occurring in it and solve for this remaining variable; if we encounter an equation which contains no new variables (hereafter called a "covered" equation) its residual will serve to correct a previously guessed variable. Whenever the number of covered equations attains the number of guessed variables (or equivalently the total number of equations attains the number of variables) a valid partition has been found and the outer iteration loop can be closed; if the number of equations ever exceeds the number of variables the sequence up to that point either contains redundant equations or is inconsistent. The cost of the sequence can be cumulated as each equation is added.

The essence of the algorithm is to carry out a depth-first search of all possible permutations, abandoning partial sequences if it is clear that completion cannot improve on the best solution so far found. The partial sequences are generated in lexicographical order, and after the addition of each equation the following tests are made:

(i) If the partial cost attains the cost of the best complete sequence so far evaluated, the last equation is removed (abandoning all completing subsequences with it in this position).

(ii) If an equation introduces no new torn variables - because it is either a covered equation or contains only one new variable - then clearly no other choice can improve on this and all completing subsequences with a different equation in this position can be abandoned.

(iii) The following "back-track rule" is applied:
Suppose at some stage equation e was added to a partial sequence of s equations, yielding a cost c_{s+1}, and that after back-tracking equation e is now added after the first r equations of this same sequence, yielding a cost $c_{r+1} \geq c_{s+1}$. Then equation e is removed (again abandoning all completing subsequences with e in this position). To demonstrate the validity of this rule we consider four cases:
(a) If e is covered by the partial sequence of r equations this case is dealt with by rule (ii) and the back-track rule is not applied.
(b) Suppose that in the original partial sequence one of the extra (s-r) equations contained no new variables not in equation e. Then this equation is either covered by the first r equations, and hence should immediately follow them, or its ouput variable must be contained in e, so that there will be one fewer torn variables if it precedes e.
(c) Suppose that each of the extra (s-r) equations contains just one new variable not in equation e. Then the increase in cost $(c_{s+1}-c_r)$ is one less than the number of new variables in equation e, which is the same as the increase obtained by placing equation e in position r+1, so that $c_{s+1} = c_{r+1}$. However in the latter case the extra (s-r) equations all introduce one new variable, and as we saw above an optimal completing subsequence will be formed by placing them immediately after equation e. But then the first (s+1) equations of this optimal sequence are the same as in the original sequence and have the same partial cost, so the total cost cannot be improved by placing equation e in position (r+1).
(d) Now suppose that one of the extra (s-r) equations contains more than one new variable not in equation e. Then if $c_{r+1} \geq c_{s+1}$ at least one of the other equations must contain no new variables not in equation e and again we have case (b). Thus the rule is proved.

It is clearly advantageous to number the equations so that the first sequence in lexicographical order has a low cost, and a good initial order is obtained by first choosing an equation with the smallest number of variables and then at each stage adding the equation which introduces fewest extra variables. In the course of this search it is easy to detect equations with identical rows in the occurrence matrix, which we call "duplicate equations". A set of duplicate equations must occur together in any sequence, since as soon as one of them occurs the remainder are automatically covered. Thus in effect the algorithm need only consider permutations of distinct rows of the occurrence matrix, with appropriate counting of duplicates.

We have already seen above that comparison of the equation and variable counts enables us to detect completion of subsets in any given sequence, and hence the closing of the outer iteration loops, but the search for the minimum tear set was not interrupted at such points. However, as soon as a subset is detected we can, if we wish, partition the sets of variables at this point and continue the lexicographical search separately within each subset. In this way the algorithm will find the sequence of irreducible subsets, and the minimum tear set within each of these.

Of course, such partitioning greatly reduces the number of permutations to be examined, and if it is to be carried out there is a strong incentive to find the partitions as early as possible. It is therefore worth "deleting" the variables already determined at each stage from the remaining equations and in the process detecting covered equations and one-variable equations. The covered equations are then added in preference to the next lexicographical equation and a check made for consistency or partitioning. If the system partitions, the one-variable equations immediately follow as one-variable subsets, and otherwise they are added to the existing partial sequence in preference to the lexicographical order (cf. Rule (ii)).

If the occurrence matrix is stored as a bit-pattern, the initial ordering, detection of duplicate equations and "deletion" of known variables are efficiently carried out by "masking" operations. It is easy to extend this algorithm to deal with a mixture of procedures and equations. We recall that a procedure cannot be executed until all its input variables are fixed, and then it yields values of all its output variables. Thus if we insert a procedure into a sequence of equations at any point we must guess any input variables not already evaluated, although the set of input variables may of course be covered by the preceding equations. Each of the resulting output variables is either a new variable or a variable previously determined; in the latter case, the difference between the two values serves as a residual for the correction of previously guessed variables. The back-track rule still applies for a procedure, and Rule (ii) applies when the whole input set is covered.

4.1 More General Cost Function

We now consider the use of a more realistic cost function in the algorithm. The difficulty in using (17) lies in the prediction of N and N_{ij} for a given subset, but a reasonable approximation is to assume that the number of iterations N_i to solve a one-variable problem by iteration is constant, and that the number of outer iterations N' is a given empirical non-decreasing function of t, the number of torn variables in the subset, yielding the form

$$c = N'(t) \sum_{i=1}^{m'} r_i N_i \quad , \tag{22}$$

where m' is the number of equations and procedures in the subset, r_i is the computational cost of evaluating a procedure or equation, N_i is unity for a procedure, explicit output variable, or residual used in the outer loop, and N_i is a given constant for an equation solved iteratively to evaluate its output variable.

It is clear that for a given computational sequence the outer loop should be closed as soon as a subset is found, so as to minimize the cost within the loop. Also, the partial cost is a non-decreasing function of t and m' as equations are added to a partial sequence. Thus Rule (i) is still valid, and Rule (ii) can be applied to a covered equation or procedure. However, Rule (ii) cannot now be applied when a one-variable equation is encountered, since there may be covered equations which would close the outer loop. In addition, if r_i is large a sequence in which this equation occurs later as a covered equation (thus reducing N_i to unity) may yield a lower cost even if t is thereby increased. This same situation also invalidates Rule (iii) since case (b) then fails.

Without the back-track rule the algorithm loses much of its power, and since this failing case is likely to be rare - and the assumptions implicit in (22) are in any case crude - there is a strong incentive to ignore it and continue to use Rule (iii). This assumed behaviour also implies, as shown in the last section, that the strategy will probably come close to minimizing the number of torn variables and that partitioning into irreducible subsets will then be profitable. Again, in view of the shortening of the algorithm, it is probably worth partitioning systematically.

4.2 Further Use of Structure

The above algorithm makes no use of the structure of the individual equations beyond the occurrence of variables in them, and possibly the cost of evaluating their residuals.

We have seen that there is little chance of building an economical assessment of conditioning or convergence rate for the outer loops into the cost function, but it may in some cases be worth carrying out a preliminary sensitivity analysis, based on the initial point as suggested by Strodiot and Toint (1976). This could then be used to choose the best output variable as each equation is considered, or even in an empirical expression for N_i, the number of one-variable iterations, once the choice is made.

It is also advantageous to choose an output variable for which the equation can be solved explicitly, without iteration, provided that the conditioning is satisfactory. More generally, instead of simply guessing all but one variable in each equation, it may be advantageous to guess variables so as to leave a subset of linear equations in several unknowns.

The logic of the algorithm described earlier is easily extended to obtain such linear subsets. It suffices to count separately the linear and nonlinear variables with the corresponding equations, so that both linear subsets and partitions may be detected. As each equation is added to the sequence it is necessary to find the minimum number of new variables which must be guessed in order that the remaining variables occur linearly. If an equation is encountered in which at most one variable occurs linearly, then all variables except the output variable must be guessed and the equation is moved to precede the current, incomplete, linear subset; other completing subsequences are abandoned. If a procedure is encountered, it is treated similarly. Of course, it is also unnecessary to consider permutations of the equations within a completed linear subset, so that there can be substantial reductions in the number of permutations examined if large linear subsets exist.

The procedure is illustrated by the example in Fig. 8, where the equations are to be added in numerical order.

Figure 8 Finding Linear Subsets

Equations	Variables 1	2	3	4	5	6	7	8	Total E	Total V	Linear E	Linear V
1		L	L	N		L	L		1	5	1	4
2	L			×			L	N	2	7	2	5
3		(L)	N					×	3	7	2	3
4	L	×	×		L	L		×	4	8	3	4
5		×	×	×	(N)				5	8	3	3
6	×			×		×		×	6	8	0	0

In the figure an N denotes a variable in the nonlinear subset, L a variable in the remaining linear subset, and × a previously determined variable; the equation and variable counts for the total (current irreducible subset) and current linear subset are given on the right. In equation 3 there is only one linear variable, so this becomes its output variable with variable 3 guessed, the equation is moved to the head of the list, and is deleted from the current linear subset. Again equation 5 contains only one nonlinear variable not previously determined, which therefore becomes its output variable, and the equation is moved to precede equation 1. This leaves equations 1, 2, 4 as a completed linear subset to determine variables 1, 6, 7. Finally, equation 6 is a covered equation, serving as an equality constraint on the values of the decision variables 3, 4, 8. If in fact equation 5 were a procedure determining variables 3 and 5 given variables 2 and 4, then the above treatment would have been the same except that the difference betwen the guessed value of variable 4 and the value given by the procedure would provide an additional residual equation and equality constraint.

This algorithm yields the sparsity structure of the linear subsets found, so that it is easy to compute the computational cost for solving the subset. For an incomplete subset, a lower bound on this cost is obtained by assuming that the completing subset coefficient matrix is the unit matrix.

Although the extension to the enumeration algorithm is logically simple, the recognition of linearly occurring variables in an equation when the values of certain variables are substituted requires some elements of an algebraic manipulation language. Once such facilities are available it becomes possible to make operation counts automatically, and perhaps to solve an equation algebraically for its output variable. Similarly differentiation becomes available, both for a preliminary sensitivity analysis and for the generation of gradients, for use in solving the equations or the nonlinear programme.

At present automatic algebraic manipulation is expensive in computing time, but in large complex engineering design problems it can assist in areas which at present require large amounts of an expensive engineer's time, particularly in formulation and consistency checking. Already interactive software systems, such as the one described by Leigh (1973), Jackson, Leigh and Sargent (1974), and Jackson (1976), are of considerable help to the designer in this phase and provide a necessary framework for use of the ideas described in this paper. The incorporation of algebraic manipulation facilities in such a system would give a significant enhancement of the aids available.

REFERENCES

1. Baker, J.J. "A Note on Multiplying Boolean Matrices", Comm. ACM, 5, 102 (1962)
2. Barkley, R.W., and Motard, R.L. "Decomposition of Nets", Chem. Eng. J., 3, 265-275 (1972)
3. Christensen, J.H., and Rudd, D.F. "Structuring Design Computations", AIChE Journal, 15, 94-100 (1969)
4. Cooper, D.C., and Whitfield, H. "ALP: AN Autocode List-Processing Language", The Computer Journal, 5, 28-32 (1962/3)
5. Duff, I.S. "A Survey of Sparse Matrix Research", AERE Harwell Report CSS28 (1976)
6. Duff, I.S., and Reid, J.K. "An Implementation of Tarjan's Algorithm for the Block Triangularization of a Matrix", AERE Harwell Report CSS29 (1976)
7. Edmonds, J. "Paths, Trees and Flowers", Can. J. Math., 17, 449-467 (1965)
8. Edmonds, J., and Johnson, E.L. "Matching: A Well-Solved Class of Integer Linear Programs", in "Combinatorial Structures and Their Applications", pp 88-92, Gordon and Breach, New York (1970)
9. Garfinkel, R.S., and Nemhauser, G.L. "Integer Programming", J. Wiley and Sons, New York (1972)
10. Harary, F. "On the Consistency of Precedence Matrices", JACM, 7, 255-259 (1960)
11. Harary, F. "A Graph Theoretic Approach to Matrix Inversion by Partitioning", Numer. Math. 4, 128-135 (1962a)
12. Harary, F. "The Determinant of the Adjacency Matrix of a Graph", SIAM Rev., 4, 202-210 (1962b)

13. Hopcroft, J.E., and Karp, R.M. "An $n^{5/2}$ Algorithm for Maximum Matchings in Bipartite Graphs", SIAM J. Comput. 2(4), 225-231 (1973)

14. Jackson, G.D.D. "Interactive Computing in Chemical Plant Design", Ph.D. Thesis, University of London (1976)

15. Lee, W., and Rudd, D.F. "On the Reordering of Recycle Calculations", AIChE Journal, 12(6), 1184-1190 (1966)

16. Leigh, M.J. "A Computer Flowsheeting Programme Incorporating Algebraic Analysis of the Problem Structure", Ph.D. Thesis, University of London (1973)

17. Leigh, M.J., Jackson, G.D.D., and Sargent, R.W.H. "SPEED-UP, a Computer-Based System for the Design of Chemical Processes", CAD 74, Fiche 15 Rows D-E; Int. Conf. and Exhib. on Computers in Engineering and Building Design (1974)

18. Mah, R.S.H. "A Constructive Algorithm for Computing the Reachability Matrix", AIChE Journal, 20(6), 1227-8 (1974)

19. Ortega, J.M., and Rheinboldt, W.C. "Iterative Solution of Nonlinear Equations in Several Variables", Academic Press, New York (1970)

20. Ponstein, J. "Self-Avoiding Paths and the Adjacency Matrix of a Graph", J. SIAM Appl. Math., 14(3), 600-609 (1966)

21. Sargent, R.W.H., and Westerberg, A.W. "SPEED-UP in Chemical Engineering Design", Trans. Instn. Chem. Engrs., 42, 190-197 (1964)

22. Steward, D.V. "On an Approach to Techniques for the Analysis of the Structure of Large Systems of Equations", SIAM Rev., 4(4), 321-342 (1962)

23. Steward, D.V. "Partitioning and Tearing Systems of Equations", J. SIAM Numer. Anal., B2(2), 345-365 (1965)

24. Strodiot, J.J., and Toint, P.L. "An Algorithm for Finding a Minimum Essential Arc Set of a Directed Graph and its Application in Solving Nonlinear Systems of Equations", Report 76/5, Dept. Mathematique, Facultés Universitaires de Namur (1976)

25. Tarjan, R. "Depth-First Search and Linear Graph Algorithms", SIAM J. Comput., 1(2), 146-160 (1972)

26. Upadhye, R.S., and E.A. Grens II. "Selection of Decompositions for Chemical Process Simulation", AIChE Journal, 21(1), 136-143 (1975)

27. Warshall, S. "A Theorem on Boolean Matrices", JACM 9, 11-12 (1962)

28. Westerberg, A.W., and Edie, F.C. Computer-Aided Design, Part 1, "Enhancing Convergence Properties by the Choice of Output Variable Assignments in the Solution of Sparse Equation Sets", Chem. Eng. J., 2, 9-16. Part 2, "An Approach to Convergence and Tearing in the Solution of Sparse Equation Sets", ibid., 17-24 (1971)

GLOBAL ERROR ESTIMATION IN ODE-SOLVERS

H.J. Stetter

I. INTRODUCTION

Consider the numerical solution of

$$(1.1) \qquad y' = f(t,y),\ y(o) = y_o; \qquad t \in [O,T],\ y(t) \in \mathbb{R}^s ,$$

by a discretization algorithm. The inherent difficulty of computing realistic bounds for the global error has been pointed out repeatedly; one of the few implemented procedures is due to Marcowitz ([1]).

In this paper we will consider error estimates. An error estimate is an approximate value of the error just as the primary solution value is an approximate value of the true solution. However, the relative accuracy requirements in this "secondary" problem are lower than for the primary problem: We are satisfied to obtain the correct sign and order of magnitude of the components. On the other hand, we are not prepared to spend much computational effort: The cost of the error estimation should remain below that of the primary computation.

We will show that it is possible to compute reliable error estimates cheaply for many algorithms for (1.1), and that this can be done concurrently with the primary computation. We will not consider round-off error effects; it will be assumed that the local round-off errors are small in relation to local discretization errors.

II. FUNDAMENTAL APPROACHES

There are (at least) two basically different approaches to the estimation of the global discretization error for (1.1):

- Compute two primary approximations from a sequence of approximations with an asymptotic behavior
- Compute the approximate defect of the primary solution and solve the error equation approximately.

A) Asymptotic estimates:

I) Richardson extrapolation: This is the classical asymptotic approach. In a fixed stepsize context one would form

$$\frac{1}{2^p - 1}\,[\eta_{2h}(t) - \eta_h(t)] \approx \varepsilon_h(t), \qquad t \in \mathbb{G}_{2h}, \tag{2.1}$$

where p ist the order of the method.

With a tolerance-controlled variable stepsize code one has to preserve coherence between the grids $\mathbb{G}_{2h}$ and $\mathbb{G}_h$ ([2],p. 73): $\mathbb{G}_h$ must be formed by halving each step in $\mathbb{G}_{2h}$; see fig. 1. Therefore, the

fig. 1

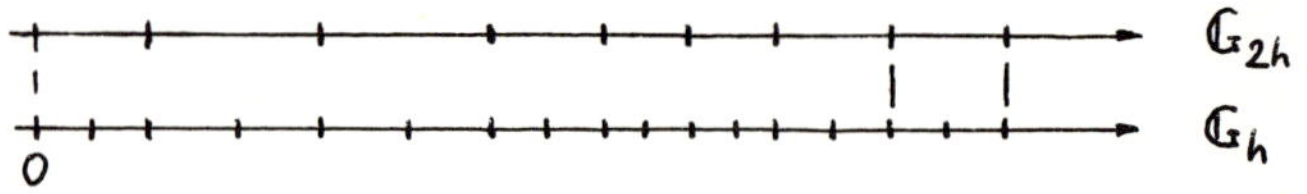

auxiliary solution η_{2h} must be computed first, under stepsize control with a tolerance $2^p \cdot$TOL if TOL would have been used ordinarily. η_h, the approximate solution proper, is then computed on the predetermined grid $\mathbb{G}_h$, without stepsize control. Obviously, the extra effort is slightly less than 50%. The computation of η_{2h} and η_h can be advanced concurrently so that ε_h is available at the same time as η_h, at $t \in \mathbb{G}_{2h}$.

Iff the basic method is a <u>one-step</u> method with <u>fixed order</u>, (2.1) holds in the described situation ([2],p. 155). The approach cannot be used with multistep methods.

This procedure has been suggested by Shampine-Watts ([3]) for global error estimation in Runge-Kutta codes, and considerable evidence for its reliability and efficiency has been compiled.

II) <u>Tolerance extrapolation</u>: The user specified tolerance parameter in an ODE-code should have this meaning (e.g. [4]): At some $t \in [0,T]$, let $\eta_{TOL}(t)$ and $\varepsilon_{TOL}(t)$ be the solution value obtained under tolerance TOL and its error resp.; then we expect, for $r > 0$,

$$\varepsilon_{r\cdot TOL}(t) \approx r\ \varepsilon_{TOL}(t). \tag{2.2}$$

If we could rely on this tolerance proportionality we could compute η_{TOL} and $\eta_{r\cdot TOL}$ $(r > 1)$ and form

$$\frac{1}{r-1}\,[\eta_{r\cdot TOL}(t) - \eta_{TOL}(t)] \approx \varepsilon_{TOL}(t). \tag{2.3}$$

But even with a perfectly "proportional" code, this would have two disadvantages:

The grids $\mathbb{G}_{r\cdot TOL}$ and $\mathbb{G}_{TOL}$ would normally not possess common points (besides $t_o = 0$). Thus one of the values in (2.3) would require interpolation which would have to preserve the tolerance proportionality.

Both computations would have to be executed under stepsize control so that, for $r = 2^p$, (2.3) would be slightly more costly than (2.1).

Good proportionality may be achieved for both one-step and multistep codes, at least at fixed orders (see [5] for a more detailed discussion). With present codes, (2.3) will often lead to less reliable error estimates than the other methods discussed in this paper.

B) Defect estimation and perturbation theory:

This has been the classical approach for theoretical purposes. There is no principal restriction to particular methods, it is equally applicable to variable order codes: Any good ODE-code should generate values of the exact solution of a perturbed problem, with a perturbation restricted by TOL. The use of perturbation theory for error estimation is the natural extension of this design goal.

I) Linearization: The effect of the defect (= perturbation) is evaluated from a variational equation. This can be done

- in the continuous realm: a defect function d(t) is formed (e.g. by interpolation and substitution into (1.1)) and the continuous error equation

$$e'(t) = f_y(t,y(t))e(t) + d(t) \tag{2.4}$$

  is evaluated, usually by discretization.

- in the discrete realm: the local error per unit step (= the discrete defect) d_n is used in the discrete error equation, e.g. (for an Adams method)

$$e_n = e_{n-1} + h_n \sum_j \beta_j \, [f_y(t_{n-j}, \eta(t_{n-j}))e_{n-j} + d_{n-j}] \, . \tag{2.5}$$

The effort consists of 3 distinct parts: (i) computation of an approximate defect, (ii) computation of (approximate) Jacobians, (iii) solution of the error equation. In the continuous approach, (i) is largely independent of the primary method and (iii) may be done by a much coarser method. For some algorithms with stepsize control (cf. sect.3) an approximate value of d_n is computed anyway.

It is (ii) which renders this approach expensive except when Jacobians have been formed in the primary computation, as in implicit methods for stiff systems (1.1). Here, the use of a variational equation is quite natural and both reliable and inexpensive; an implementation and experimental results have been reported by Prothero and Robinson

([10]).

II) <u>Defect correction</u>: We use this term as defined in [6]. To understand the modification needed for error estimation let us look at the general iterative procedure first:

We assume that a continuous problem

(2.6) $\quad Fx = 0$

with true solution x^* has been discretized into

(2.7) $\quad \tilde{\phi}\xi = 0$

and that problems of the type $\tilde{\phi}\xi = \lambda$ can be readily solved numerically. Furthermore, a defect defining function ϕ assigns defects to grid functions.

fig. 2

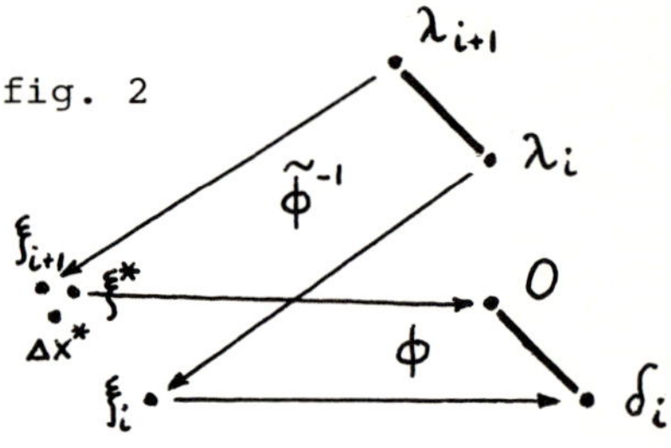

Then the typical loop of an <u>iterative</u> defect correction (IDeC) algorithm is*) (see fig. 2):

$$(2.8) \quad \begin{cases} (\xi_i \text{ was previously computed from } \tilde{\phi}\xi_i = \lambda_i,) \\ \text{form the defect } \delta_i := \phi\xi_i, \\ \text{compute } \xi_{i+1} \text{ from } \quad \tilde{\phi}\xi_{i+1} = \lambda_i - \delta_i =: \lambda_{i+1} \end{cases}$$

If $\tilde{\phi}^{-1}[\tilde{\phi} - \phi]$ is contracting, $\{\xi_i\} \to \xi^*$ where $\phi\xi^* = 0$ and $\|\xi_i - \Delta x^*\| \to \|\xi^* - \Delta x^*\|$ (Δ is some natural discretization operator).

For <u>error estimation</u>, we want to use a cheaper method than $\tilde{\phi}$ for the only iteration to be executed; hence we identify the solution ξ of (2.7) with an intermediate ξ_i of an IDeC with a method $\approx\!\!\phi$. $\lambda_i = \tilde{\tilde{\phi}}\xi_i$ has now to be formed a posteriori, otherwise we simply replace $\tilde{\phi}$ by $\tilde{\tilde{\phi}}$ in (2.8); see fig. 3:

$$(2.9) \quad \begin{cases} \text{form } \lambda := \tilde{\tilde{\phi}}\xi, \\ \text{form } \delta := \phi\xi, \\ \text{compute } \bar{\xi} \text{ from } \tilde{\tilde{\phi}}\bar{\xi} = \lambda - \delta. \end{cases}$$

fig. 3

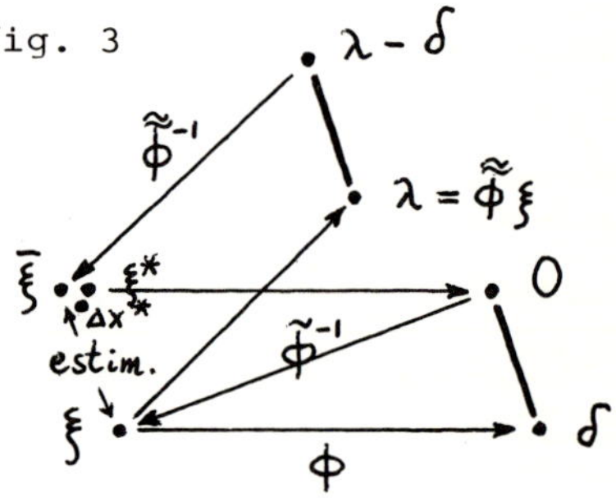

Obviously,

$$\bar{\xi} - \xi^* = \tilde{\tilde{\phi}}^{-1}[\tilde{\tilde{\phi}} - \phi]\xi - \tilde{\tilde{\phi}}^{-1}[\tilde{\tilde{\phi}} - \phi]\xi^*$$

and our error estimate $\xi - \bar{\xi}$ satisfies

---

*) There exists a different, non-equivalent version of IDeC which we shall not discuss here; cf. [6].

$$(2.10)\qquad \xi - \bar{\xi} = \underbrace{(\xi - \Delta x^*)}_{\text{true error}} - \underbrace{(\bar{\xi} - \xi^*)}_{\text{contracted error of } \xi} - \underbrace{(\xi^* - \Delta x^*)}_{\text{effect of defect estimation error}}\ ;$$

it is a good representation of the true error if the last two terms in (2.10) are comparatively small. This requires sufficient contractivity of $\tilde{\tilde{\phi}}^{-1}[\tilde{\tilde{\phi}} - \phi]$ and a reasonable quality of the defect estimator ϕ.

An asymptotic analysis shows: If ϕ produces a lowest order correct value of the true local error per unit step then $\tilde{\tilde{\phi}}$ need only be first order convergent to be "consistent" with ϕ. For (1.1), this points to the explicit Euler method as the natural choice, or to the implicit Euler method if (1.1) is stiff.

III. ALGORITHMIC REALIZATION OF THE DEFECT CORRECTION APPROACH

Assume the primary method to be "non-stiff". According to (2.9) we may proceed thus for $n = 1,2,\ldots$:

After we have produced the primary value η_n at the next gridpoint t_n, with a step $h_n := t_n - t_{n-1}$,

(i) we form

$$(3.1)\qquad \lambda_n := \frac{\eta_n - \eta_{n-1}}{h_n} - f(t_{n-1}, \eta_{n-1});$$

(ii) we compute (or retrieve from the primary computation) a lowest order correct value δ_n of the local error per unit step*)

$$(3.2)\qquad \delta_n = \frac{\eta_n - z_n(t_n)}{h_n}\,(1 + O(h_n));$$

(iii) we "solve"

$$(3.3)\qquad \frac{\bar{\eta}_n - \bar{\eta}_{n-1}}{h_n} - f(t_{n-1}, \bar{\eta}_{n-1}) = \lambda_n - \delta_n;$$

(iv) we obtain our error estimate from

$$(3.4)\qquad \varepsilon_n := \eta_n - \bar{\eta}_n$$

or rather, to avoid the cancellation in (3.4), from (cf. (3.1)/(3.3))

$$(3.5)\qquad \varepsilon_n := \varepsilon_{n-1} + h_n[(f(t_{n-1}, \eta_{n-1}) - f(t_{n-1}, \bar{\eta}_{n-1})) + \delta_n].$$

---

*) the local true solution z_n satisfies $z_n'(t) = f(t, z_n(t)), z_n(t_{n-1}) = \eta_{n-1}$.

In this form, the error estimation procedure may be attached to virtually all discretization algorithms for (1.1). The primary computation is not affected at all by the attachment of (3.1) - (3.5); it can make full use of its stepsize and order control mechanisms which determine its efficiency, reliability, and robustness. (Only if the primary method $\tilde{\phi}$ has a large absolute stability region and happens to be used near its boundary, the secondary computation may be subject to weak instability and some recourse may have to be taken.)

The actual computation of λ_n will normally be different from (3.1), since $\eta_n = \eta_{n-1} + h_n[..]$ in most methods. The evaluation of $f(t_{n-1}, \eta_{n-1})$ will nearly always occur in the primary code and not constitute an extra effort. The only unavoidable extra f-evaluation is the one in (3.3).

Let us now consider the computation of δ_n. One would expect that, in a code with local stepsize control, an estimate of the local error would have to be computed anyway. But most of the recent ODE-codes make use of what has been called "local extrapolation": In each step, they determine a more accurate value $\eta_n^{(1)}$ and a less accurate value $\eta_n^{(o)}$. While the approximate local error $\eta_n^{(o)} - \eta_n^{(1)}$ of $\eta_n^{(o)}$ is used for stepsize control purposes, the solution is continued with $\eta_n^{(1)}$. In these codes with "fictitious local error estimates", δ_n is thus not available from the primary computation; in fact, its evaluation constitutes the major effort in (3.1) - (3.5). However, as δ_n is needed only after the completion of the primary step, all the information from this step is available.

Nonetheless, in RK-codes with local extrapolation, the computation of δ_n is likely to be rather expensive if not information from more than one step is used or special RK-methods ("embedded triplets") have been constructed.

For Adams PECE-codes with local extrapolation, δ_n can be computed with <u>one</u> additional f-evaluation for arbitrary step ratios and local orders (section 4). Therèfore, the defect correction approach seems particularly well suited for this class of powerful and flexible codes for which no other reliable method of global error estimation is presently available.

If the primary method is stiff and applied to a stiff problem (1.1), the implicit Euler method has to be used in (3.1)/(3.3): Then

$$(3.6)\qquad \begin{aligned} \lambda_n &:= \frac{\eta_n - \eta_{n-1}}{h_n} - f(t_n, \eta_n) , \\ \frac{\bar{\eta}_n - \bar{\eta}_{n-1}}{h_n} &- h_n f(t_n, \bar{\eta}_n) = \lambda_n - \delta_n \end{aligned}$$

The triangular decomposition of $I - h_n f_y(t_n, \eta_n)$ should be available from the primary computation. A quasi-Newton iteration with this Jacobian, starting from η_n, should produce the solution $\bar{\eta}_n$ of (3.6) in few iterations. Thus the implicit Euler step (3.6) should again require considerably less computational effort than the same step in the primary solution process.

IV. ANALYSIS AND IMPLEMENTATION FOR A VARIABLE ORDER, VARIABLE STEP ADAMS PECE-CODE

Assume that information from k_n previous points has been used in stepping from t_{n-1} to t_n and that a fictitious local error estimate necessitates the a posteriori computation of δ_n.

The local error per unit step δ_n consists of two parts

$$(4.1) \qquad \delta_n = M_n + N_n$$

where M_n contains the remaining effect of the predictor and N_n is the local error per unit step of the corrector. Both terms are of the same order asymptotically. We have to determine "lowest order correct" values of M_n and N_n. (In using the $O(h_n)$-symbol in the variable step context, we assume that $h_{n-1},\dots,h_{n-k_n+1}$ are fixed multiples of h_n.)

The details of the following analysis have been elaborated in [8], where it has been shown that the lowest order term $\tilde{M}_n$ of M_n satisfies

$$(4.2) \qquad \tilde{M}_n := g_{k_n}(f_n^p - f_n)(1 + O(h_n));$$

here f_n^p and f_n are the known predicted and corrected values of f at t_n and g_{k_n} depends only on the local grid structure. The lowest order term $\tilde{N}_k$ of N_n, however, cannot be expressed in terms of quantities from the primary computation; otherwise, we could eliminate $\tilde{M}_n$ and $\tilde{N}_n$ from η_n and gain a more accurate value at no extra cost.

Let P_n^p and P_n be the polynomials of degree k_n interpolating $f_{n-k_n}, \dots, f_{n-1}$ and f_n^p resp. f_n; let $t_{n-\frac{1}{2}} := t_{n-1} + h_n/2$. From

$$(4.3) \qquad \eta_{n-\frac{1}{2}} := \eta_n + \int_{t_{n-1}}^{t_{n-\frac{1}{2}}} P_n^p(t)\,dt \quad \text{and} \quad \eta'_{n-\frac{1}{2}} := P_n(t_{n-\frac{1}{2}})$$

we form the defect

$$(4.4) \qquad d_{n-\frac{1}{2}} := \eta'_{n-\frac{1}{2}} - f(t_{n-\frac{1}{2}}, \eta_{n-\frac{1}{2}}).$$

Then, as shown in [8],

$$(4.5) \qquad \tilde{N}_n = \beta_{k_n}^{\frac{1}{2}} \, g^*_{k_n} \, d_{n-\frac{1}{2}} \, (1+O(h_n))$$

where $\beta_k^{\frac{1}{2}}$ and g^*_k depend only on the local grid structure.

The cost of evaluating (4.1) via (4.2) - (4.5) is not high: g_{k_n} and $g^*_{k_n}$ are formed within the primary computation, $\beta_{k_n}^{1/2}$ requires a few arithmetic operations. A subroutine which evaluates (4.3) for arbitrary t_{n-r}, $r \in (0,1)$, from primary information with arithmetic operations only, is part of a flexible Adams code anyhow (see, e.g., [7]). The only new f-evaluation occurs in (4.4).

The procedure (4.1) - (4.5) has an important advantage over other possibilities (like interpolation of η_n-values and formation of continuous defect): Since it uses precisely the information which was used or generated in the current primary step it automatically adapts to the events in the primary computation. Thus, no special starting provision is necessary, the local order of the δ_n-estimate is increased or decreased with that of the current step, and - most important - no information deemed unreliable by the primary order control mechanism is used in the defect computation.

Experimental comparisons of computed δ_n-values and true local errors per unit step over various test problems (with known local solutions), wide ranges of tolerances, and long ranges of integration have established the validity and reliability of this procedure; see [8].

There remains one question: Effective codes do not adapt the grid to specified output points t_{out} but form $\eta(t_{out})$ by interpolation, see (4.3) and [7]; so, how should we define $\varepsilon(t_{out})$?

Considering (3.3) and (3.5), <u>linear</u> interpolation is the only reasonable choice. But this implies the following interpolation for η in (t_{n-1}, t_n):

$$(4.6) \qquad \eta_n(t) = z_n(t) + (t - t_{n-1})\delta_n .$$

Instead of this ideal interpolation, (4.3) produces (see [8])

$$(4.7) \qquad \tilde{\eta}_n(t) = z_n(t) + h_n\left(W_n\left(\frac{t-t_{n-1}}{h_n}\right) M_n + W^*_n\left(\frac{t-t_{n-1}}{h_n}\right) N_n\right)(1+O(h_n)),$$

where W_n and W^*_n are polynomials depending on k_n and the local gridstructure, with $W_n(0) = W^*_n(0) = 0$, $W_n(1) = W^*_n(1) = 1$.

fig. 4

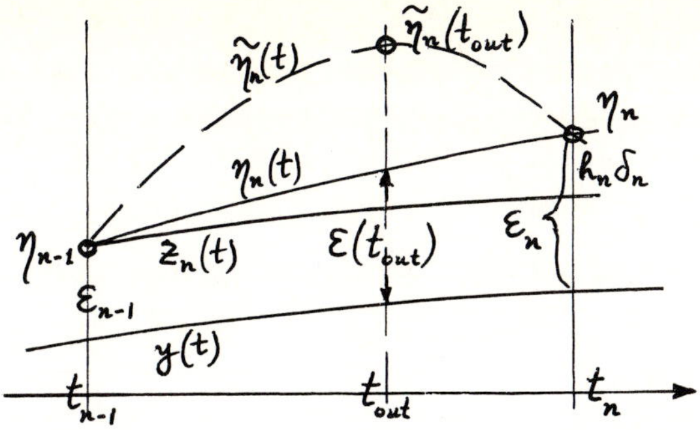

The discrepancy between (4.6) and (4.7) is tolerable if (componentwise) $|\varepsilon_n| \gg h_n \max(|\delta_n|, |M_n|, |N_n|)$, cf. fig.4. But in stable ODEs often $|\varepsilon_n| \approx h_n|\delta_n|$; then the major part of the global error at t_{out} may be "interpolation error" $\tilde{\eta}_n(t_{out}) - \eta_n(t_{out})$ which is not reflected by our computed $\varepsilon(t_{out})$.

There is an elegant solution: We may use the computed approximations to M_n and N_n (see (4.2) and (4.5)) to transform $\tilde{\eta}_n(t_{out})$ into $\eta_n(t_{out})$. The computation of the W_n- and W_n^*-values requires only a slight extension of a recursion used in the interpolation code for $\tilde{\eta}_n$. This transformation actually has two advantages: $\eta_n(t)$ constitutes a smoother interpolation (fig. 4), and the computed error estimate between grid points becomes more reliable.

V. RELIABILITY OF GLOBAL ERROR ESTIMATES

In section 1, we have requested that our estimates should correctly represent the order of magnitude and the sign of each component. Thus, in the situation of fig. 5 we will require our estimate to lie between the dashed curves.

fig. 6

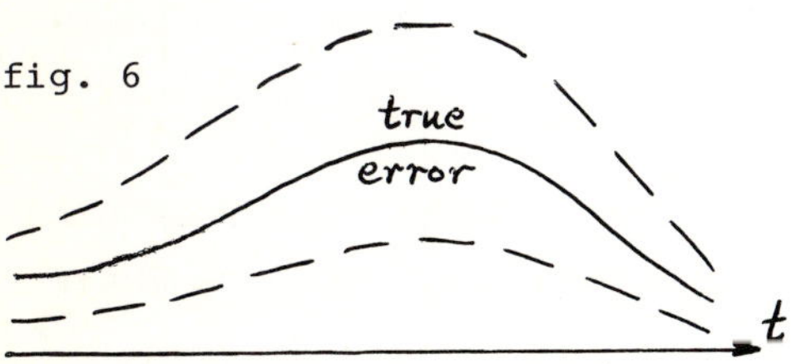

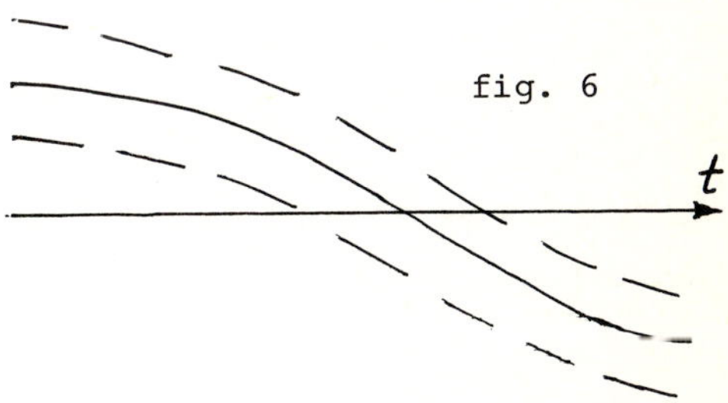

However, this requirement has to be modified in several natural ways. If a component of the global error changes sign, an estimate as indicated in fig. 6 is fully acceptable although it may locally misrepresent both the sign and the size of the error. Furthermore, it has to be taken into account that a global error estimate is essentially a numerical solution of (2.4). If (1.1) is unstable in the analytic sense, the accuracy of any numerical solution of (2.4) will deteriorate for increasing t. Similarly, if (1.1) describes a nonlinear oscillation, the primary and the secondary approximation may contain phase errors growing with t. Thus, the values at given times may deviate much more than the trajectories, both in the primary and the secondary computation.

The tightness of the "fit" of the error estimate should improve with tighter tolerances in the primary computation. This improvement will come to an end, and the computed error will gradually become meaningless as local round-off errors begin to outweigh local discretization errors.

A simple but costly way to assess the reliability of a global error estimate for a reasonably proportional code (cf. section 2 A II) is the following: We rerun the problem, with active error estimation, at tolerance level r.TOL so that we obtain, at some t

$$\eta_{TOL}(t),\ \varepsilon_{TOL}(t),\ \eta_{r.TOL}(t),\ \varepsilon_{r.TOL}(t).$$

Ideally, the signs of ε_{TOL}, $\varepsilon_{r.TOL}$, and $\eta_{r.TOL} - \eta_{TOL}$ should be the same and their sizes should relate like $1 : r : (r-1)$, $r \gg 1$ (cf. (2.2),(2.3)). The occurence of this situation would thus support the reliability of the estimate. On the other hand, a serious deviation would indicate that - locally at least - the reliability of the estimate was impaired (e.g. by a change in sign).

This check has proved effective in actual computation (see [9]).

VI. CONCLUSIONS

The procedure outlined in sections 3 and 4 has been fully implemented "on top" of the code STEP by Shampine-Gordon [7]. Extensive tests have fully confirmed the expectations; their results have been reported in [9].

These results, and those of [3] and others, indicate that reliable and inexpensive global error estimates can indeed be provided by library routines for initial value problems in ODEs. At the present state of the art the following approaches seem most appropriate:

one-step, fixed order codes	Richardson extrapolation on
fictitiuos local error estimates	coherent grids (section 2AI)

any type (non-stiff) correct local error estimates Adams PC, variable order fictitious local error estimates	defect correction with explicit Euler integrator (section 3)
"stiff" solvers correct local error estimates	solution of error equation (section 2BI) or defect correction with implicit Euler integration (section 3)

Reliable global error estimates would considerably enhance the safety and efficiency of present ODE-codes: Unduly wide tolerances lead to inaccurate results while too stringent tolerances waste computer time. If a user could "switch on" global error estimation as an option, he could determine the appropriate value of TOL for his problem in a preliminary run and then use the code most efficiently in subsequent production runs.

It is hoped that in a few years time this option will be a matter of course in library routines.

REFERENCES

[1] U. Marcowitz: Fehlerabschätzung bei Anfangswertaufgaben für Systeme gewöhnlicher Differentialgleichungen, Num. Math. 24 (1975) 249-275.

[2] H.J. Stetter: Analysis of Discretization Methods for Ordinary Differential Equations, Springer-Verlag, Berlin-Heidelberg-New York, 1973.

[3] L.F. Shampine, H.A. Watts: Global Error Estimation for Ordinary Differential Equations, TOMS 2 (1976) 172-186.

[4] H.J. Stetter: Considerations concerning a Theory for ODE-Solvers, to appear

[5] H.J. Stetter: Tolerance Proportionality in ODE-codes, to appear.

[6] H.J. Stetter: The Defect Correction Principle and Discretization Methods, to appear in Num. Math.

[7] L.F. Shampine, M.K. Gordon: Computer Solution of Ordinary Differential Equations, W.H. Freeman & Co., San Francisco, 1975.

[8] H.J. Stetter: Interpolation and Error Estimation in Adams PC-Codes, to appear.

[9] H.J. Stetter: Global Error Estimation in Adams PC-codes, to appear.

[10] A. Prothero, A. Robinson: Global Error Estimates to Stiff Systems of ODE, submitted paper, Dundee Conference, June 1977.

ISOJACOBIC CROSSWIND DIFFERENCING

Eugene L. Wachspress *

1. Isoparametric Geometry.

The Imperial College TEACH program [1] solves the steady-state Navier-Stokes equations in primitive variables and conservative form over an Eulerian grid of rectangles in either Cartesian or cylindrical coordinates by iterating over interlocking velocity and pressure grids (Fig. 1a) to consistency of five coupled linear elliptic systems derived with such features as hybrid differencing [3] , a two-equation turbulence model, and logarithmic wall-boundary velocity profiles. As a first step in generalizing TEACH geometry, an isoparametric type [4] element was introduced. Although the "box integration" discretization scheme in TEACH is retained in the new TURF program, the isoparametric geometry requires doubling the number of unknowns (Fig. 1b).

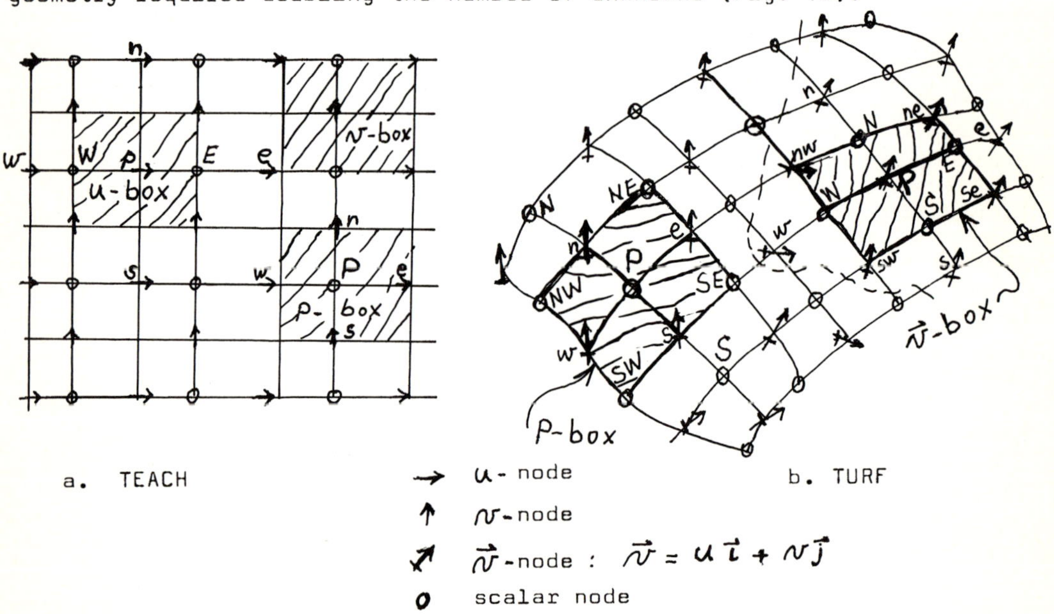

Fig. 1. TEACH and TURF Elements.

Each interior scalar node in TURF is determined as the origin of an isoparametric transformation having a constant Jacobian on the coordinate axes which pass through the neighboring velocity nodes. We now describe the coordinate system about scalar node P in Fig. 1b. The most general isoparametric coordinate system (ξ , η) is related to (x,y) by quadratic equations:

---

* Knolls Atomic Power Laboratory, operated for the U.S. Energy Research and Development Administration by the General Electric Company, Contract No. EY-76-C-12-0052.

$$x = a_0' + a_1'\xi + a_2'\eta + a_3'\xi^2 + a_4'\eta^2 + a_5'\xi\eta \,, \tag{1.1}$$

$$y = b_0' + b_1'\xi + b_2'\eta + b_3'\xi^2 + b_4'\eta^2 + b_5'\xi\eta \,. \tag{1.2}$$

The coefficients are determined uniquely when lines (e;w) and (n;s) are not parallel by the following conditions:

$$\begin{array}{l} (\xi,\eta)_w = (-1,0) \quad (\xi,\eta)_e = (1,0) \quad (\xi,\eta)_s = (0,-1) \\ (\xi,\eta)_n = (0,1) \quad x_{\xi\eta} = y_{\xi\eta} = 0 \\ J(\xi,\eta) = x_\xi y_\eta - x_\eta y_\xi = J_0 \text{ (a constant) on } \xi = 0 \text{ and on } \eta = 0. \end{array} \tag{1.3}$$

This choice is particularly well suited for TURF discretization. Conditions (1.3) reduce (1.1) and (1.2) to

$$\begin{array}{l} x = a_0 + a_1(\xi + \alpha\eta^2) + a_2(\eta + \beta\xi^2) \\ y = b_0 + b_1(\xi + \alpha\eta^2) + b_2(\eta + \beta\xi^2), \end{array} \tag{1.4}$$

with

$$\begin{array}{l} \alpha = \dfrac{(y_n - y_s)(x_e + x_w - x_n - x_s) - (x_n - x_s)(y_e + y_w - y_n - y_s)}{(y_e - y_w)(x_n - x_s) - (x_e - x_w)(y_n - y_s)} \\ \beta = \dfrac{(y_e - y_w)(x_e + x_w - x_n - x_s) - (x_e - x_w)(y_e + y_w - y_n - y_s)}{(y_e - y_w)(x_n - x_s) - (x_e - x_w)(y_n - y_s)} \end{array} \tag{1.5}$$

$$\begin{array}{l} a_0 = \tfrac{1}{2}\left[(x_n + x_s) - \alpha(x_e - x_w)\right] \qquad b_0 = \tfrac{1}{2}\left[(y_e + y_w) - \beta(y_n - y_s)\right] \\ a_1 = \tfrac{1}{2}(x_e - x_w) \quad b_1 = \tfrac{1}{2}(y_e - y_w) \quad a_2 = \tfrac{1}{2}(x_n - x_s) \quad b_2 = \tfrac{1}{2}(y_n - y_s) \,. \end{array} \tag{1.6}$$

Point P is the isoparametric origin $(\xi,\eta) = (0,0)$ with (x,y) coordinates (a_0,b_0). The location of boundary scalar nodes are problem data, but a transformation of the type given in (1.4) still applies. For example, if P is on a North boundary and (w;P;e) is not a straight line, then [for the case where $(y_e + y_w - 2y_p) \neq 0$]

$$\begin{array}{l} (a_0,b_0) = (x_P,y_P); \quad (a_1,b_1) = \left(\dfrac{x_e - x_w}{2}, \dfrac{y_e - y_w}{2}\right); \\ (a_2,b_2) = \left(b_2\gamma, \dfrac{b_1(x_s - a_0) - a_1(y_s - b_0)}{a_1 - b_1\gamma}\right) \\ \text{with } \gamma \text{ defined as } \quad \gamma \equiv (x_e + x_w - 2x_P)/(y_e + y_w - 2y_P); \\ \alpha = (a_2 + x_s - a_0)/a_1 \qquad \beta = (x_e + x_w - 2x_P)/2a_2 \,. \end{array} \tag{1.7}$$

The Jacobians of an isoparametric transformation satisfy

$$J\left(\frac{x,y}{\xi,\eta}\right) = \begin{bmatrix} x_\xi & x_\eta \\ y_\xi & y_\eta \end{bmatrix}; \quad J\left(\frac{\xi,\eta}{x,y}\right) = \begin{bmatrix} \xi_x & \xi_y \\ \eta_x & \eta_y \end{bmatrix} \quad J = \det J\left(\frac{x,y}{\xi,\eta}\right) = x_\xi y_\eta - x_\eta y_\xi \,. \tag{1.8}$$

Elements are restricted so that $J \neq 0$ in any element, and we choose coordinates so that J is positive throughout.

We note that $J^{-1}(\frac{x,y}{\xi,\eta}) = J(\frac{\xi,\eta}{x,y})$. Hence,

$$\frac{1}{J}\begin{bmatrix} y_\eta & -x_\eta \\ -y_\xi & x_\xi \end{bmatrix} = \begin{bmatrix} \xi_x & \xi_y \\ \eta_x & \eta_y \end{bmatrix} . \tag{1.9}$$

Several vector and tensor relationships are useful in the discretization of the Navier-Stokes equations. We define the covariant and contravariant base vectors

$$\vec{g}_1 = x_\xi \vec{i} + y_\xi \vec{j} , \quad \vec{g}_2 = x_\eta \vec{i} + y_\eta \vec{j} , \quad \vec{g}^1 = (y_\eta \vec{i} - x_\eta \vec{j})/J , \quad \vec{g}^2 = (-y_\xi \vec{i} + x_\xi \vec{j})/J . \tag{1.10}$$

Then

$$\begin{aligned} &\vec{g}_i \cdot \vec{g}^j = 0 \text{ when } i \neq j \qquad \nabla f \cdot \vec{g}_1 = f_\xi \qquad \nabla f \cdot \vec{g}_2 = f_\eta \\ &\phantom{\vec{g}_i \cdot \vec{g}^j} = 1 \text{ when } i = j \\ &\qquad \nabla \cdot \vec{g}_1 = J_\xi / J \qquad \nabla \cdot \vec{g}_2 = J_\eta / J \\ &\nabla f = f_\xi \vec{g}^1 + f_\eta \vec{g}^2 \qquad \nabla \vec{g}_1 = \vec{g}^1 (x_{\xi\xi} \vec{i} + y_{\xi\xi} \vec{j}) \\ &\qquad \nabla \vec{g}_2 = \vec{g}^2 (x_{\eta\eta} \vec{i} + y_{\eta\eta} \vec{j}). \end{aligned} \tag{1.11}$$

Velocity $\vec{v} = u\vec{i} + v\vec{j}$ may be expressed in isoparametric coordinates for interpolation in each of the four directions from a velocity node to its neighboring velocity nodes. The coordinate system is in general different for each direction. We have

$$\vec{v} = u\vec{i} + v\vec{j} = v^1 \vec{g}_1 + v^2 \vec{g}_2 , \tag{1.12}$$

where

$$v^1 = \vec{v} \cdot \vec{g}^1 = u(\vec{i} \cdot \vec{g}^1) + v(\vec{j} \cdot \vec{g}^1) = (y_\eta u - x_\eta v)/J$$

and

$$v^2 = \vec{v} \cdot \vec{g}^2 = u(\vec{i} \cdot \vec{g}^2) + v(\vec{j} \cdot \vec{g}^2) = (x_\xi v - y_\xi u)/J \tag{1.13}$$

Moreover,

$$u = x_\xi v^1 + x_\eta v^2 \quad \text{and} \quad v = y_\xi v^1 + y_\eta v^2 . \tag{1.14}$$

Velocities interpolated from velocity nodes to scalar nodes are used for approximating integrals along isoparametric arcs on which the Jacobian determinants are constant. We refer to such interpolation as "isojacobic."

For our special isoparametric transformation, we find that

$$\nabla \vec{v} = v^1_\xi \vec{g}^1 \vec{g}_1 + v^1_\eta \vec{g}^2 \vec{g}_1 + v^2_\xi \vec{g}^1 \vec{g}_2 + v^2_\eta \vec{g}^2 \vec{g}_2 + v^1 \vec{g}^1 (x_{\xi\xi} \vec{i} + y_{\xi\xi} \vec{j}) + v^2 \vec{g}^2 (x_{\eta\eta} \vec{i} + y_{\eta\eta} \vec{j}). \tag{1.15}$$

Working out numerical examples in this coordinate system helps one gain familiarity **with** the tensor formalism in analysis of fluid flow.

2. Crosswind Differencing.

Computational stability and accuracy are strongly dependent on velocity interpolation from velocity-node values to scalar-node values. The interpolation must account for the relative importance of transport and diffusion. The TEACH hybrid interpolation is essentially equivalent to use of upwind differences for cell Reynold's numbers greater than two (transport dominated) and central differences otherwise (diffusion dominated). This scheme [1,3] may be viewed at scalar node P in Fig. 1a as the valuesthere being obtained by approximate solution of the two-point boundary-value-problems

$$G_P^1\, u - \mu_P u_x = \text{constant}, \quad x_w \le x \le x_e \qquad \text{(u-interpolation)}$$
$$u(x_w) = u_w\,, \quad u(x_e) = u_e$$

where ρ is density and μ viscosity at P, and G_P^1 is the latest estimate of ρu_P.

$$G_P^2 v - \mu_P v_y = \text{constant}, \quad y_s \le y \le y_n$$
$$v(y_s) = v_s\,, \quad v(y_n) = v_n \qquad \text{(v-interpolation at P)}$$
$$G_P^2 \text{ - latest estimate of } \rho v_P.$$

The error introduced by this approximation is equivalent to that resulting from introduction of "artificial viscosity" [2], and is most pronounced when the flow is at 45° to the grid.

The differential equations for the interpolation may be obtained by choosing as constant each of the components of the dyad $\rho\vec{v}\vec{v} - \mu\nabla\vec{v}$ appearing in the Navier-Stokes equations. Ambiguitioo not in TEACH arise in TURF where u_P could be interpolated as above from the $\vec{i}\vec{i}$ component of the dyad or from the $\vec{j}\vec{i}$ component $\rho v_P u - \mu u_y$ with boundary values of u_n and u_s. To eliminate this ambiguity and to develop an interpolation procedure of higher order accuracy we consider a two-dimensional interpolation based on the partial differential equation

$$\nabla \cdot \left[(\rho\vec{v})_{old}\vec{v} - \mu\nabla\vec{v}\right] = \vec{0} \tag{2.1}$$

and boundary conditions at all four velocity neighbors of P for both the u- and the v-interpolation. We use the $\vec{i}$-component of (2.1) as the governing equation for u-interpolation and the $\vec{j}$-component for v-interpolation. The interpolated value of u at P is the value at P of the function U obtained by solving the boundary-value-problem

$$G_P^1\, U_x - \mu U_{xx} + G_P^2\, U_y - \mu U_{yy} = 0\,, \tag{2.2}$$
$$U_w = u(x_w,y_w),\quad U_e = u(x_e,y_e),\quad U_n = u(x_n,y_n),\quad U_s = u(x_s,y_s).$$

Solution of (2.2) requires more than point boundary conditions. However, we may choose a particularly simple form for U that satisfies (2.2) and the point boundary conditions:

$$U(x,y) = X(x) + Y(y) = c_1 \exp(G^1 x/\mu) + c_2 \exp(G^2 y/\mu) + c_3(G^1 y - G^2 x) + c_4 \,. \tag{2.3}$$

Choice of this additive form rather than a multiplicative form, $U = YX$, simplifies the algebra. The resulting interpolation formulas seem to account for the competition between transport and diffusion effects. Let $x_e - x_w = h$ and $y_n - y_s = k$. Define

$$r_1 = (\rho u h/2\mu)_P \quad \text{and} \quad r_2 = (\rho v k/2\mu)_P \,. \tag{2.4}$$

The values for the c_i in (2.3) are uniquely determined by the boundary conditions in (2.2), and the value of U at P is:

$$u_P = \frac{u_e\left[T(r_1) - r_1\right] + u_w\left[T(r_1) + r_1\right] + u_n\left[T(r_2) - r_2\right] + u_s\left[T(r_2) + r_2\right]}{2\left[T(r_1) + T(r_2)\right]} \tag{2.5}$$

where $T(r) = r \coth(r/2)$. In hybrid differencing analogous to that in TEACH one approximates $T(r)$ by $|r|$ when $r \gtrsim 2$ and by 2 when $r < 2$. Interpolation with (2.5) allows the wind to blow in any direction. For example, if both r_1 and r_2 are greater than two, then

$$u_P = \frac{r_1 u_w + r_2 u_s}{r_1 + r_2} \,. \tag{2.6}$$

If in (2.6) $r_1 = r_2$, then $u_P = \frac{1}{2}(u_w + u_s)$. Flow at 45° to the grid lines no longer aggravates artificial viscosity.

A more accurate representation of the hyperbolic cotangents further reduces the error, and in TURF we choose

$$\begin{aligned} T(r) &= 2 + \frac{r^2}{6} - \frac{r^4}{360} \quad \text{when} \quad |r| \le 3.8786 \\ &= |r| \quad \text{when} \quad |r| > 3.8786 \,. \end{aligned} \tag{2.7}$$

Interpolation formulas for the TURF isoparametric elements may be derived in similar fashion. The algebra is more complex, and we replace all variable coefficients in the linearized differential equations by their values at P. We define w'_{ij} by

$$\rho\vec{v}\vec{v} - \mu\nabla\vec{v} = w'_{11}\vec{g}_1\vec{g}_1 + w'_{21}\vec{g}_2\vec{g}_1 + w'_{12}\vec{g}_1\vec{g}_2 + w'_{22}\vec{g}_2\vec{g}_2 . \tag{2.8}$$

From (1.10)-(1.15) we obtain

$$\begin{aligned} w'_{11} &= \rho v^1 v^1 - \mu\vec{g}^1 \cdot \nabla\vec{v} \cdot \vec{g}^1 \\ &= \rho v^1 v^1 - \mu\Big[v^1_\xi \vec{g}^1 \cdot \vec{g}^1 + v^1_\eta \vec{g}^1 \cdot \vec{g}^2 + v^1(\vec{g}^1 \cdot \vec{g}^1)(x_{\xi\xi}\vec{i} + y_{\xi\xi}\vec{j}) \cdot \vec{g}^1 \\ &\qquad + v^2(\vec{g}^1 \cdot \vec{g}^2)(x_{\eta\eta}\vec{i} + y_{\eta\eta}\vec{j}) \cdot \vec{g}^1\Big] \\ &= \rho v^1 v^1 - \frac{\mu}{J^2}\Big[(y_\eta^2 + x_\eta^2)\Big(v^1_\xi + v^1 \frac{y_\eta x_{\eta\eta} - x_\eta y_{\eta\eta}}{J}\Big) \\ &\qquad - (x_\xi x_\eta + y_\xi y_\eta)\Big(v^1_\eta + v^2 \frac{y_\eta x_{\eta\eta} - x_\eta y_{\eta\eta}}{J}\Big)\Big] \end{aligned} \tag{2.9}$$

We note that $J_\xi = (x_\xi y_\eta - x_\eta y_\xi)_\xi = x_{\xi\xi} y_\eta - y_{\xi\xi} x_\eta = 0$ at P and that $y_\eta x_{\eta\eta} - x_\eta y_{\eta\eta} = (b_2 + 2b_1\alpha\eta)2a_1\alpha - (a_2 + 2a_1\alpha\eta)2b_1\alpha = 2J_0\alpha$.
Now let w_{ij} be the value of w'_{ij} with variable coefficients replaced by values at P. Then

$$w_{11} = (\rho v^1)_P v^1 - \frac{\mu}{J_0^2}\Big[(a_2^2 + b_2^2)v^1_\xi - (a_1 a_2 + b_1 b_2)(v^1_\eta + 2\alpha v^2)\Big]. \tag{2.10}$$

We define

$$G_1 \equiv \rho v^1_{\text{latest known value at P}} \;;\; G_2 \equiv \rho v^2_{\text{latest known value at P}}$$

$$\mu_1 \equiv \mu \frac{x_\eta^2 + y_\eta^2}{J^2}\Big|_P = \mu\frac{a_2^2 + b_2^2}{J_0^2}, \quad \mu_2 \equiv \mu\frac{x_\xi^2 + y_\xi^2}{J^2}\Big|_P = \mu\frac{a_1^2 + b_1^2}{J_0^2} \tag{2.11}$$

$$\sigma \equiv \mu\frac{x_\xi x_\eta + y_\xi y_\eta}{J^2}\Big|_P = \mu\frac{a_1 a_2 + b_1 b_2}{J_0^2}$$

Then

$$w_{11} = G_1 v^1 - \mu_1 v^1_\xi + (2\alpha v^2 + v^1_\eta)\sigma. \tag{2.12a}$$

Similar analysis yields

$$w_{21} = G_2 v^1 - \mu_2(2\alpha v^2 + v^1_\eta) + \sigma v^1_\xi \tag{2.12b}$$

$$w_{12} = G_1 v^2 - \mu_1(2\beta v^1 + v^2_\xi) + \sigma v^2_\eta \tag{2.12c}$$

$$w_{22} = G_2 v^2 - \mu_2 v^2_\eta + \sigma(2\beta v^1 + v^2_\xi). \tag{2.12d}$$

We now consider $\nabla\cdot(w_{ij}\vec{g}_i\vec{g}_j)$. For any dyad $w\vec{a}\vec{b}$, we note that

$$\nabla\cdot w\vec{a}\vec{b} = (\nabla\cdot w\vec{a})\vec{b} + w\vec{a}\cdot\nabla\vec{b}. \tag{2.13}$$

Thus,

$$\nabla\cdot(w_{11}\vec{g}_1 + w_{21}\vec{g}_2)\vec{g}_1 = \big[\nabla\cdot(w_{11}\vec{g}_1 + w_{21}\vec{g}_2)\big]\vec{g}_1 + (w_{11}\vec{g}_1 + w_{21}\vec{g}_2)\cdot\nabla\vec{g}_1. \tag{2.14}$$

Using (1.11), we reduce the last term to

$$(w_{11}\vec{g}_1 + w_{21}\vec{g}_2)\cdot\vec{g}^1(x_{\xi\xi}\vec{i} + y_{\xi\xi}\vec{j}) = w_{11}(x_{\xi\xi}\vec{i} + y_{\xi\xi}\vec{j}). \tag{2.15}$$

Further use of (1.11) yields

$$\begin{aligned} \nabla\cdot w\vec{g}_1 &= w\,\nabla\cdot\vec{g}_1 + \nabla w\cdot\vec{g}_1 = w(J_\xi/J) + w_\xi = w_\xi, \text{ and} \\ \nabla\cdot w\vec{g}_2 &= w\,\nabla\cdot\vec{g}_2 + \nabla w\cdot\vec{g}_2 = w(J_\eta/J) + w_\eta = w_\eta, \end{aligned} \tag{2.16}$$

where we have set J_ξ and J_η to zero, consistent with our previous approximations in terms of values at interpolation point P.
It follows that

$$\nabla\cdot(w_{11}\vec{g}_1 + w_{21}\vec{g}_2) = (w_{11})_\xi + (w_{21})_\eta \tag{2.17}$$

and that

$$\nabla\cdot(\rho\vec{v}\vec{v} - \mu\nabla v) = \big[(w_{11})_\xi + (w_{21})_\eta\big]\vec{g}_1 + \big[(w_{12})_\xi + (w_{22})_\eta\big]\vec{g}_2 + \vec{s} \tag{2.18}$$

where $\vec{s} = w_{11}(x_{\xi\xi}\vec{i} + y_{\xi\xi}\vec{j}) + w_{22}(x_{\eta\eta}\vec{i} + y_{\eta\eta}\vec{j})$ can be expressed in terms of $\vec{g}_1$ and $\vec{g}_2$:

$$\vec{s} = 2\alpha w_{22}\vec{g}_1 + 2\beta w_{11}\vec{g}_2. \tag{2.19}$$

For the v^1 interpolation we seek a function that satisfies

$$(w_{11})_\xi + (w_{21})_\eta + 2\alpha w_{22} = 0 \tag{2.20a}$$

and boundary conditions on v^1 at points n, s, e and w. For the v^2 interpolation we seek a function that satisfies

$$(w_{12})_\xi + (w_{22})_\eta + 2\beta w_{11} = 0 \tag{2.20b}$$

and boundary conditions on v^2 at n, s, e and w. The interpolation is greatly simplified if we drop the terms with α and β in (2.20). Actually, $\nabla\cdot(\rho\vec{v}\vec{v} - \mu\nabla\vec{v})$ is in general nonzero in the neighborhood of P. Suppose $\nabla\cdot(\rho\vec{v}\vec{v} - \mu\nabla\vec{v}) = f_1\vec{g}_1 + f_2\vec{g}_2$ and that f_1 and f_2 could be estimated from latest values around P. Then (2.20a) could be replaced by

$$(w_{11})_\xi + (w_{21})_\eta + 2\alpha w_{22} = f_1. \tag{2.21}$$

Substituting (2.12) into (2.21), we would obtain

$$\begin{aligned} G_1 v^1_\xi - \mu_1 v^1_{\xi\xi} + 2\alpha\sigma v^2_\xi + \sigma v^1_{\eta\xi} + G_2 v^1_\eta - \mu_2 v^1_{\eta\eta} - 2\alpha\mu_2 v^2_\eta + \sigma v^1_{\xi\eta} \\ + 2\alpha\left[G_2 v^2 - \mu_2 v^2_\eta + \sigma(2\beta v^1 + v^2_\xi)\right] = f_1 . \end{aligned} \tag{2.22}$$

We could then define in terms of latest values

$$f^1 = f_1 - 4\alpha\left[\tfrac{1}{2}G_2 v^2 - \mu_2 v^2_\eta + \sigma(\beta v^1 + v^2_\xi)\right] \tag{2.23}$$

and use as the differential equation to be satisfied by the v^1 interpolation function:

$$G_1 v^1_\xi - \mu_1 v^1_{\xi\xi} + 2\sigma v^1_{\xi\eta} + G_2 v^1_\eta - \mu_2 v^1_{\eta\eta} = f^1 . \tag{2.24}$$

Use of (2.20a) with the terms in α and β dropped is equivalent to setting f^1 to zero in (2.24). The key to successful interpolation is appropriate accounting for the relative importance of transport and diffusion-type phenomena, and we shall presently see that this is accomplished satisfactorily with $f^1 = 0$ in (2.24). The additional computation in estimating and using a nonzero f^1 does not seem warranted in our application.

Referring to (2.2), we note that the additive form of the interpolation function eliminates the $v^1_{\xi\eta}$ term in (2.24) and reduces (2.24) to the same form as (2.2). If we now define

$$r_1 \equiv G_1/2\mu_1 \qquad \text{and} \qquad r_2 \equiv G_2/2\mu_2 , \tag{2.25}$$

then (2.5) generalizes to

$$v^i = \frac{\mu_1\{v^i_e[T(r_1) - r_1] + v^i_w[T(r_1) + r_1]\} + \mu_2\{[v^i_n T(r_2) - r_2] + v^i_s[T(r_2) + r_2]\}}{2\left[\mu_1 T(r_1) + \mu_2 T(r_2)\right]} \tag{2.26}$$

$$i = 1,2.$$

Thus, the crosswind interpolation of (2.5) for rectangular geometry may be generalized to isoparametric geometry through use of the directional viscosities defined by (2.11). Equation (2.26) has all the features previously described for (2.5) and thus allows for the competition between transport and diffusion momentum transfer.

3. Momentum Integrals.

To illustrate the discretization procedure in TURF, we consider the momentum equations. The Navier-Stokes equation is integrated over each $\vec{v}$-box (Fig. 1). The x- and y-components of the resulting vector equation are the "velocity equations". Values for u and v are computed at each velocity node. The interpolation is expressed in terms of v^1 and v^2. Eqs. (1.12)-(1.14) are used to relate u,v to v^1,v^2:

$\begin{bmatrix} v^1 \\ v^2 \end{bmatrix} = \frac{1}{J_0} T \begin{bmatrix} u \\ v \end{bmatrix}$ with matrix T defined for each velocity-node neighbor of scalar node P in terms of the coordinates at P by

$$T_e = \begin{bmatrix} b_2 & -a_2 \\ -(b_1+2b_2\beta) & (a_1+2a_2\beta) \end{bmatrix} \qquad T_w = \begin{bmatrix} b_2 & -a_2 \\ (2b_2\beta-b_1) & (a_1-2a_2\beta) \end{bmatrix}$$

$$T_n = \begin{bmatrix} (b_2+2b_1\alpha) & -(a_2+2a_1\alpha) \\ -b_1 & a_1 \end{bmatrix} \qquad T_s = \begin{bmatrix} (b_2-2b_1\alpha) & (2a_1\alpha-a_2) \\ -b_1 & a_1 \end{bmatrix} \tag{3.1}$$

and at scalar node P itself by

$$T_P = \begin{bmatrix} b_2 & -a_2 \\ -b_1 & a_1 \end{bmatrix}. \tag{3.2}$$

To illustrate application of (2.26), (3.1) and (3.2), we evaluate

$$\iint_{\vec{v}\text{-box}} dxdy\, \nabla\cdot(\rho\vec{v}\vec{v} - \mu\nabla\vec{v}) = \oint_{\partial(\vec{v}\text{-box})} \vec{n}\cdot(\rho\vec{v}\vec{v} - \mu\nabla\vec{v})ds\ . \tag{3.3}$$

Referring to the $\vec{v}$-box in Fig. 1, we evaluate the contribution to (3.3) from integration over side (se;E;ne). The other contributions may be evaluated in similar fashion. The x-component is obtained by dotting (3.3) on the right with $\vec{i}$:

$$\oint_{(se;E;ne)} \vec{n}ds\cdot(\rho\vec{v}\vec{v} - \mu\nabla\vec{v})\cdot\vec{i} = \int_{-1}^{1} d\eta\ J(\rho v^1 u - \mu\vec{g}^1\cdot\nabla u). \tag{3.4}$$

Expanding $\vec{g}^1$ in terms of $\vec{g}_1$ and $\vec{g}_2$, we obtain with the aid of (1.11):

$$\vec{g}^1\cdot\nabla u = \frac{x_\eta^2 + y_\eta^2}{J^2}u_\xi - \frac{x_\xi x_\eta + y_\xi y_\eta}{J^2}u_\eta\ . \tag{3.5}$$

Referring to (2.11), we find that

$$(\mu g^1 \cdot \nabla u)_E = (\mu_1 u_\xi - \sigma u_\eta)_E. \tag{3.6}$$

Central differences suffice for approximating u_ξ and u_η here:

$$u_\xi \doteq (u_e - u_p)/2 \qquad \text{and} \qquad u_\eta \doteq (u_{ne} - u_{se})/2 \;. \tag{3.7}$$

Let the integral in (3.4) be denoted by $M_x(se;E;ne)$. Then

$$M_x(se;E;ne) = \left[2G_{1E}u_E - \mu_1(u_e - u_p) + \sigma(u_{ne} - u_{se})\right] J_E. \tag{3.8}$$

Similarly, the contribution to the y-momentum equation is

$$M_y(se;E;ne) = \left[2G_{1E}v_E - \mu_1(v_e - v_p) + \sigma(v_{ne} - v_{se})\right] J_E. \tag{3.9}$$

Eq. (2.26) yields at each scalar node the coefficients for velocity interpolation at the node which we now write in the form

$$v^i_E = c_e v^i_e + c_p v^i_p + c_{ne} v^i_{ne} + c_{se} v^i_{se} \;, \; i = 1,2. \tag{3.10}$$

A TURF iteration cycle involves successive solution of velocity, pressure, and two turbulence equations. During the velocity iteration the coefficients in (3.10) remain fixed. They are updated at the end of the cycle. We must determine u_E and v_E in (3.8) and (3.9) as functions of the velocities at neighboring nodes, using the transformations in (3.1) and (3.2) together with interpolation formula (3.10). For this purpose we define the matrices

$$I = \begin{bmatrix} 1 & 0 \\ 0 & 1 \end{bmatrix} \qquad M_1 = \begin{bmatrix} -a_1 b_1 & (a_1)^2 \\ -(b_1)^2 & a_1 b_1 \end{bmatrix} \qquad M_2 = \begin{bmatrix} -a_2 b_2 & (a_2)^2 \\ -(b_2)^2 & a_2 b_2 \end{bmatrix} \tag{3.11a}$$

$$C_e = c_e(I + \frac{2\beta}{J} M_2), \quad C_p = c_p(I - \frac{2\beta}{J} M_2), \quad C_{ne} = c_{ne}(I - \frac{2\alpha}{J} M_1)$$
$$C_{se} = c_{se}(I + \frac{2\alpha}{J} M_1). \tag{3.11b}$$

It can then be shown that

$$\underline{v}_E = \begin{bmatrix} u_E \\ v_E \end{bmatrix} = C_e \underline{v}_e + C_p \underline{v}_p + C_{ne} \underline{v}_{ne} + C_{se} \underline{v}_{se} \;. \tag{3.12}$$

The contribution to the vector momentum equation (3.3) given in (3.8) and (3.9) may then be expressed as

$$\underline{M}(se;E;ne) = J_E\Big[(\mu_1 I + 2\rho v^1 C_p)\underline{v}_p - (\mu_1 I - 2\rho v^1 C_e)\underline{v}_e$$
$$+ (\sigma I + 2\rho v^1 C_{ne})\underline{v}_{ne} - (\sigma I - 2\rho v^1 C_{se})\underline{v}_{se} \;. \tag{3.13}$$

The values for μ_1, ρ, σ, v^1 and the elements in the C matrices in (3.13) are latest values at node E. Thus, the coefficients in (3.13) do not change during the velocity iteration.

4. Concluding Remarks.

A possible generalization from the rectangular elements in TEACH to isoparametric elements has been described. The one-dimensional hybrid differencing in TEACH appears inadequate for the generalized geometry, and a more sophisticated two-dimensional hybrid differencing was introduced. Computational complexity is increased considerably when TEACH is generalized in this manner. Twice as many unknowns must be found each iteration. Suitable combinations of geometry parameters may be stored in advance for use in the flow computation. The formalism may be applied to time-dependent flow and to three-space-dimension analysis. Such computation will not be attempted until it has been demonstrated that the two-dimensional steady-state TURF program performs well.

5. References.

1. Gosman, A. D. and Pun, W. M., Lecture notes for a course on "Calculation of recirculating flows," Imperial College Report No. HTS/74/2 (1973).

2. Roache, P. J., Computational Fluid Dynamics, Hermosa Publishers, Albuquerque, New Mexico (1976).

3. Spalding, D. B., "A novel finite difference formulation for differential expressions involving both first and second derivatives," International Journal for Numerical Methods in Engineering, Vol. 4, pp. 551-559 (1972).

4. Zienkiewicz, O. C., The Finite Element Method in Engineering Science (2nd ed.), McGraw-Hill, New York (1971).

Vol. 460: O. Loos, Jordan Pairs. XVI, 218 pages. 1975.

Vol. 461: Computational Mechanics. Proceedings 1974. Edited by J. T. Oden. VII, 328 pages. 1975.

Vol. 462: P. Gérardin, Construction de Séries Discrètes p-adiques. »Sur les séries discrètes non ramifiées des groupes réductifs déployés p-adiques«. III, 180 pages. 1975.

Vol. 463: H.-H. Kuo, Gaussian Measures in Banach Spaces. VI, 224 pages. 1975.

Vol. 464: C. Rockland, Hypoellipticity and Eigenvalue Asymptotics. III, 171 pages. 1975.

Vol. 465: Séminaire de Probabilités IX. Proceedings 1973/74. Edité par P. A. Meyer. IV, 589 pages. 1975.

Vol. 466: Non-Commutative Harmonic Analysis. Proceedings 1974. Edited by J. Carmona, J. Dixmier and M. Vergne. VI, 231 pages. 1975.

Vol. 467: M. R. Essén, The Cos $\pi\lambda$ Theorem. With a paper by Christer Borell. VII, 112 pages. 1975.

Vol. 468: Dynamical Systems – Warwick 1974. Proceedings 1973/74. Edited by A. Manning. X, 405 pages. 1975.

Vol. 469: E. Binz, Continuous Convergence on C(X). IX, 140 pages. 1975.

Vol. 470: R. Bowen, Equilibrium States and the Ergodic Theory of Anosov Diffeomorphisms. III, 108 pages. 1975.

Vol. 471: R. S. Hamilton, Harmonic Maps of Manifolds with Boundary. III, 168 pages. 1975.

Vol. 472: Probability-Winter School. Proceedings 1975. Edited by Z. Ciesielski, K. Urbanik, and W. A. Woyczyński. VI, 283 pages. 1975.

Vol. 473: D. Burghelea, R. Lashof, and M. Rothenberg, Groups of Automorphisms of Manifolds. (with an appendix by E. Pedersen) VII, 156 pages. 1975.

Vol. 474: Séminaire Pierre Lelong (Analyse) Année 1973/74. Edité par P. Lelong. VI, 182 pages. 1975.

Vol. 475: Répartition Modulo 1. Actes du Colloque de Marseille-Luminy, 4 au 7 Juin 1974. Edité par G. Rauzy. V, 258 pages. 1975. 1975.

Vol. 476: Modular Functions of One Variable IV. Proceedings 1972. Edited by B. J. Birch and W. Kuyk. V, 151 pages. 1975.

Vol. 477: Optimization and Optimal Control. Proceedings 1974. Edited by R. Bulirsch, W. Oettli, and J. Stoer. VII, 294 pages. 1975.

Vol. 478: G. Schober, Univalent Functions – Selected Topics. V, 200 pages. 1975.

Vol. 479: S. D. Fisher and J. W. Jerome, Minimum Norm Extremals in Function Spaces. With Applications to Classical and Modern Analysis. VIII, 209 pages. 1975.

Vol. 480: X. M. Fernique, J. P. Conze et J. Gani, Ecole d'Eté de Probabilités de Saint-Flour IV–1974. Edité par P.-L. Hennequin. XI, 293 pages. 1975.

Vol. 481: M. de Guzmán, Differentiation of Integrals in R^n. XII, 226 pages. 1975.

Vol. 482: Fonctions de Plusieurs Variables Complexes II. Séminaire François Norguet 1974–1975. IX, 367 pages. 1975.

Vol. 483: R. D. M. Accola, Riemann Surfaces, Theta Functions, and Abelian Automorphisms Groups. III, 105 pages. 1975.

Vol. 484: Differential Topology and Geometry. Proceedings 1974. Edited by G. P. Joubert, R. P. Moussu, and R. H. Roussarie. IX, 287 pages. 1975.

Vol. 485: J. Diestel, Geometry of Banach Spaces – Selected Topics. XI, 282 pages. 1975.

Vol. 486: S. Stratila and D. Voiculescu, Representations of AF-Algebras and of the Group U (∞). IX, 169 pages. 1975.

Vol. 487: H. M. Reimann und T. Rychener, Funktionen beschränkter mittlerer Oszillation. VI, 141 Seiten. 1975.

Vol. 488: Representations of Algebras, Ottawa 1974. Proceedings 1974. Edited by V. Dlab and P. Gabriel. XII, 378 pages. 1975.

Vol. 489: J. Bair and R. Fourneau, Etude Géométrique des Espaces Vectoriels. Une Introduction. VII, 185 pages. 1975.

Vol. 490: The Geometry of Metric and Linear Spaces. Proceedings 1974. Edited by L. M. Kelly. X, 244 pages. 1975.

Vol. 491: K. A. Broughan, Invariants for Real-Generated Uniform Topological and Algebraic Categories. X, 197 pages. 1975.

Vol. 492: Infinitary Logic: In Memoriam Carol Karp. Edited by D. W. Kueker. VI, 206 pages. 1975.

Vol. 493: F. W. Kamber and P. Tondeur, Foliated Bundles and Characteristic Classes. XIII, 208 pages. 1975.

Vol. 494: A Cornea and G. Licea. Order and Potential Resolvent Families of Kernels. IV, 154 pages. 1975.

Vol. 495: A. Kerber, Representations of Permutation Groups II. V, 175 pages. 1975.

Vol. 496: L. H. Hodgkin and V. P. Snaith, Topics in K-Theory. Two Independent Contributions. III, 294 pages. 1975.

Vol. 497: Analyse Harmonique sur les Groupes de Lie. Proceedings 1973–75. Edité par P. Eymard et al. VI, 710 pages. 1975.

Vol. 498: Model Theory and Algebra. A Memorial Tribute to Abraham Robinson. Edited by D. H. Saracino and V. B. Weispfenning. X, 463 pages. 1975.

Vol. 499: Logic Conference, Kiel 1974. Proceedings. Edited by G. H. Müller, A. Oberschelp, and K. Potthoff. V, 651 pages 1975.

Vol. 500: Proof Theory Symposion, Kiel 1974. Proceedings. Edited by J. Diller and G. H. Müller. VIII, 383 pages. 1975.

Vol. 501: Spline Functions, Karlsruhe 1975. Proceedings. Edited by K. Böhmer, G. Meinardus, and W. Schempp. VI, 421 pages. 1976.

Vol. 502: János Galambos, Representations of Real Numbers by Infinite Series. VI, 146 pages. 1976.

Vol. 503: Applications of Methods of Functional Analysis to Problems in Mechanics. Proceedings 1975. Edited by P. Germain and B. Nayroles. XIX, 531 pages. 1976.

Vol. 504: S. Lang and H. F. Trotter, Frobenius Distributions in GL_2-Extensions. III, 274 pages. 1976.

Vol. 505: Advances in Complex Function Theory. Proceedings 1973/74. Edited by W. E. Kirwan and L. Zalcman. VIII, 203 pages. 1976.

Vol. 506: Numerical Analysis, Dundee 1975. Proceedings. Edited by G. A. Watson. X, 201 pages. 1976.

Vol. 507: M. C. Reed, Abstract Non-Linear Wave Equations. VI, 128 pages. 1976.

Vol. 508: E. Seneta, Regularly Varying Functions. V, 112 pages. 1976.

Vol. 509: D. E. Blair, Contact Manifolds in Riemannian Geometry. VI, 146 pages. 1976.

Vol. 510: V. Poènaru, Singularités C^∞ en Présence de Symétrie. V, 174 pages. 1976.

Vol. 511: Séminaire de Probabilités X. Proceedings 1974/75. Edité par P. A. Meyer. VI, 593 pages. 1976.

Vol. 512: Spaces of Analytic Functions, Kristiansand, Norway 1975. Proceedings. Edited by O. B. Bekken, B. K. Øksendal, and A. Stray. VIII, 204 pages. 1976.

Vol. 513: R. B. Warfield, Jr. Nilpotent Groups. VIII, 115 pages. 1976.

Vol. 514: Séminaire Bourbaki vol. 1974/75. Exposés 453 – 470. IV, 276 pages. 1976.

Vol. 515: Bäcklund Transformations. Nashville, Tennessee 1974. Proceedings. Edited by R. M. Miura. VIII, 295 pages. 1976.

Vol. 516: M. L. Silverstein, Boundary Theory for Symmetric Markov Processes. XVI, 314 pages. 1976.

Vol. 517: S. Glasner, Proximal Flows. VIII, 153 pages. 1976.

Vol. 518: Séminaire de Théorie du Potentiel, Proceedings Paris 1972–1974. Edité par F. Hirsch et G. Mokobodzki. VI, 275 pages. 1976.

Vol. 519: J. Schmets, Espaces de Fonctions Continues. XII, 150 pages. 1976.

Vol. 520: R. H. Farrell, Techniques of Multivariate Calculation. X, 337 pages. 1976.

Vol. 521: G. Cherlin, Model Theoretic Algebra – Selected Topics. IV, 234 pages. 1976.

Vol. 522: C. O. Bloom and N. D. Kazarinoff, Short Wave Radiation Problems in Inhomogeneous Media: Asymptotic Solutions. V. 104 pages. 1976.

Vol. 523: S. A. Albeverio and R. J. Høegh-Krohn, Mathematical Theory of Feynman Path Integrals. IV, 139 pages. 1976.

Vol. 524: Séminaire Pierre Lelong (Analyse) Année 1974/75. Edité par P. Lelong. V, 222 pages. 1976.

Vol. 525: Structural Stability, the Theory of Catastrophes, and Applications in the Sciences. Proceedings 1975. Edited by P. Hilton. VI, 408 pages. 1976.

Vol. 526: Probability in Banach Spaces. Proceedings 1975. Edited by A. Beck. VI, 290 pages. 1976.

Vol. 527: M. Denker, Ch. Grillenberger, and K. Sigmund, Ergodic Theory on Compact Spaces. IV, 360 pages. 1976.

Vol. 528: J. E. Humphreys, Ordinary and Modular Representations of Chevalley Groups. III, 127 pages. 1976.

Vol. 529: J. Grandell, Doubly Stochastic Poisson Processes. X, 234 pages. 1976.

Vol. 530: S. S. Gelbart, Weil's Representation and the Spectrum of the Metaplectic Group. VII, 140 pages. 1976.

Vol. 531: Y.-C. Wong, The Topology of Uniform Convergence on Order-Bounded Sets. VI, 163 pages. 1976.

Vol. 532: Théorie Ergodique. Proceedings 1973/1974. Edité par J.-P. Conze and M. S. Keane. VIII, 227 pages. 1976.

Vol. 533: F. R. Cohen, T. J. Lada, and J. P. May, The Homology of Iterated Loop Spaces. IX, 490 pages. 1976.

Vol. 534: C. Preston, Random Fields. V, 200 pages. 1976.

Vol. 535: Singularités d'Applications Differentiables. Plans-sur-Bex. 1975. Edité par O. Burlet et F. Ronga. V, 253 pages. 1976.

Vol. 536: W. M. Schmidt, Equations over Finite Fields. An Elementary Approach. IX, 267 pages. 1976.

Vol. 537: Set Theory and Hierarchy Theory. Bierutowice, Poland 1975. A Memorial Tribute to Andrzej Mostowski. Edited by W. Marek, M. Srebrny and A. Zarach. XIII, 345 pages. 1976.

Vol. 538: G. Fischer, Complex Analytic Geometry. VII, 201 pages. 1976.

Vol. 539: A. Badrikian, J. F. C. Kingman et J. Kuelbs, Ecole d'Eté de Probabilités de Saint Flour V-1975. Edité par P.-L. Hennequin. IX, 314 pages. 1976.

Vol. 540: Categorical Topology, Proceedings 1975. Edited by E. Binz and H. Herrlich. XV, 719 pages. 1976.

Vol. 541: Measure Theory, Oberwolfach 1975. Proceedings. Edited by A. Bellow and D. Kölzow. XIV, 430 pages. 1976.

Vol. 542: D. A. Edwards and H. M. Hastings, Čech and Steenrod Homotopy Theories with Applications to Geometric Topology. VII, 296 pages. 1976.

Vol. 543: Nonlinear Operators and the Calculus of Variations, Bruxelles 1975. Edited by J. P. Gossez, E. J. Lami Dozo, J. Mawhin, and L. Waelbroeck, VII, 237 pages. 1976.

Vol. 544: Robert P. Langlands, On the Functional Equations Satisfied by Eisenstein Series. VII, 337 pages. 1976.

Vol. 545: Noncommutative Ring Theory. Kent State 1975. Edited by J. H. Cozzens and F. L. Sandomierski. V, 212 pages. 1976.

Vol. 546: K. Mahler, Lectures on Transcendental Numbers. Edited and Completed by B. Diviš and W. J. Le Veque. XXI, 254 pages. 1976.

Vol. 547: A. Mukherjea and N. A. Tserpes, Measures on Topological Semigroups: Convolution Products and Random Walks. V, 197 pages. 1976.

Vol. 548: D. A. Hejhal, The Selberg Trace Formula for PSL (2,ℝ). Volume I. VI, 516 pages. 1976.

Vol. 549: Brauer Groups, Evanston 1975. Proceedings. Edited by D. Zelinsky. V, 187 pages. 1976.

Vol. 550: Proceedings of the Third Japan – USSR Symposium on Probability Theory. Edited by G. Maruyama and J. V. Prokhorov. VI, 722 pages. 1976.

Vol. 551: Algebraic K-Theory, Evanston 1976. Proceedings. Edited by M. R. Stein. XI, 409 pages. 1976.

Vol. 552: C. G. Gibson, K. Wirthmüller, A. A. du Plessis and E. J. N. Looijenga. Topological Stability of Smooth Mappings. V, 155 pages. 1976.

Vol. 553: M. Petrich, Categories of Algebraic Systems. Vector and Projective Spaces, Semigroups, Rings and Lattices. VIII, 217 pages. 1976.

Vol. 554: J. D. H. Smith, Mal'cev Varieties. VIII, 158 pages. 1976.

Vol. 555: M. Ishida, The Genus Fields of Algebraic Number Fields. VII, 116 pages. 1976.

Vol. 556: Approximation Theory. Bonn 1976. Proceedings. Edited by R. Schaback and K. Scherer. VII, 466 pages. 1976.

Vol. 557: W. Iberkleid and T. Petrie, Smooth S^1 Manifolds. III, 163 pages. 1976.

Vol. 558: B. Weisfeiler, On Construction and Identification of Graphs. XIV, 237 pages. 1976.

Vol. 559: J.-P. Caubet, Le Mouvement Brownien Relativiste. IX, 212 pages. 1976.

Vol. 560: Combinatorial Mathematics, IV, Proceedings 1975. Edited by L. R. A. Casse and W. D. Wallis. VII, 249 pages. 1976.

Vol. 561: Function Theoretic Methods for Partial Differential Equations. Darmstadt 1976. Proceedings. Edited by V. E. Meister, N. Weck and W. L. Wendland. XVIII, 520 pages. 1976.

Vol. 562: R. W. Goodman, Nilpotent Lie Groups: Structure and Applications to Analysis. X, 210 pages. 1976.

Vol. 563: Séminaire de Théorie du Potentiel. Paris, No. 2. Proceedings 1975–1976. Edited by F. Hirsch and G. Mokobodzki. VI, 292 pages. 1976.

Vol. 564: Ordinary and Partial Differential Equations, Dundee 1976. Proceedings. Edited by W. N. Everitt and B. D. Sleeman. XVIII, 551 pages. 1976.

Vol. 565: Turbulence and Navier Stokes Equations. Proceedings 1975. Edited by R. Temam. IX, 194 pages. 1976.

Vol. 566: Empirical Distributions and Processes. Oberwolfach 1976. Proceedings. Edited by P. Gaenssler and P. Révész. VII, 146 pages. 1976.

Vol. 567: Séminaire Bourbaki vol. 1975/76. Exposés 471–488. IV, 303 pages. 1977.

Vol. 568: R. E. Gaines and J. L. Mawhin, Coincidence Degree, and Nonlinear Differential Equations. V, 262 pages. 1977.

Vol. 569: Cohomologie Etale SGA 4½. Séminaire de Géométrie Algébrique du Bois-Marie. Edité par P. Deligne. V, 312 pages. 1977.

Vol. 570: Differential Geometrical Methods in Mathematical Physics, Bonn 1975. Proceedings. Edited by K. Bleuler and A. Reetz. VIII, 576 pages. 1977.

Vol. 571: Constructive Theory of Functions of Several Variables, Oberwolfach 1976. Proceedings. Edited by W. Schempp and K. Zeller. VI, 290 pages. 1977.

Vol. 572: Sparse Matrix Techniques, Copenhagen 1976. Edited by V. A. Barker. V, 184 pages. 1977.

Vol. 573: Group Theory, Canberra 1975. Proceedings. Edited by R. A. Bryce, J. Cossey and M. F. Newman. VII, 146 pages. 1977.

Vol. 574: J. Moldestad, Computations in Higher Types. IV, 203 pages. 1977.

Vol. 575: K-Theory and Operator Algebras, Athens, Georgia 1975. Edited by B. B. Morrel and I. M. Singer. VI, 191 pages. 1977.

Vol. 576: V. S. Varadarajan, Harmonic Analysis on Real Reductive Groups. VI, 521 pages. 1977.

Vol. 577: J. P. May, E_∞ Ring Spaces and E_∞ Ring Spectra. IV, 268 pages. 1977.

Vol. 578: Séminaire Pierre Lelong (Analyse) Année 1975/76. Edité par P. Lelong. VI, 327 pages. 1977.

Vol. 579: Combinatoire et Représentation du Groupe Symétrique, Strasbourg 1976. Proceedings 1976. Edité par D. Foata. IV, 339 pages. 1977.